PEASANTS AND THE ART OF FARMING

A CHAYANOVIAN MANIFESTO

Jan Douwe van der Ploeg

AGRARIAN CHANGE AND PEASANT STUDIES SERIES

Fernwood Publishing
Halifax and Winnipeg

Practical Action Publishing Ltd
The Schumacher Centre, Bourton on Dunsmore, Rugby, Warwickshire, CV23 9QZ, UK
www.practicalactionpublishing.org

First edition published by Fernwood Publishing, Canada, 2013
This edition published by Practical Action Publishing Ltd, 2014

© Jan Douwe van der Ploeg

The right of the author to be identified as author of the work has been asserted under sections 77 and 78 of the Copyright Designs and Patents Act 1988.

All rights reserved. No part of this publication may be reprinted or reproduced or utilized in any form or by any electronic, mechanical, or other means, now known or hereafter invented, including photocopying and recording, or in any information storage or retrieval system, without the written permission of the publishers.

Product or corporate names may be trademarks or registered trademarks, and are used only for identification and explanation without intent to infringe.

A catalogue record for this book is available from the British Library.
A catalogue record for this book has been requested from the Library of Congress.

ISBN 978-1-85339-876-6 Hardback
ISBN 978-1-85339-877-3 Paperback
ISBN 978-1-78044-876-3 Library Ebook
ISBN 978-1-78044-877-0 Ebook

Citation: Douwe van der Ploeg, J. (2013) Peasants and the Art of Farming, Rugby, UK: Practical Action Publishing, <http://dx.doi.org/10.3362/9781780448763>

Since 1974, Practical Action Publishing has published and disseminated books and information in support of international development work throughout the world. Practical Action Publishing is a trading name of Practical Action Publishing Ltd (Company Reg. No. 1159018), the wholly owned publishing company of Practical Action. Practical Action Publishing trades only in support of its parent charity objectives and any profits are covenanted back to Practical Action (Charity Reg. No. 247257, Group VAT Registration No. 880 9924 76).

The views and opinions in this publication are those of the author and do not represent those of Practical Action Publishing Ltd or its parent charity Practical Action. Reasonable efforts have been made to publish reliable data and information, but the authors and publisher cannot assume responsibility for the validity of all materials or for the consequences of their use.

Cover design by John van der Woude

Contents

ICAS Agrarian Change and Peasant Studies Series v
Acknowledgements viii
Preface ix

1 Peasants and Social Transformations 1
    The Political Relevance of Peasant Theory 11
    Peasant Agriculture and Capitalism 15
    What Makes Chayanov a "Genius"? 17
    A Proof of Pedigree 20

2 The Two Main Balances Identified by Chayanov 23
    The Peasant Unit of Production: No Wages, No Capital 24
    The Labour-Consumer Balance 33
    The Political Relevance of the Labour-Consumer Balance 34
    The Scientific Relevance of the Labour-Consumer Balance 35
    The Balance of Utility and Drudgery 38
    On "Subjective Evaluation" 42
    Self-Exploitation 44

3 A Wider Array of Interacting Balances 48
    The Balance between People and Living Nature 48
    The Balance of Production and Reproduction 54
    The Balance of Internal and External Resources 56
    The Balance of Autonomy and Dependence 60
    The Balance of Scale and Intensity
    (and the Emergence of Farming Styles) 63
    Fighting for Progress in an Adverse Environment 66
    By Way of Synthesis: The Peasant Farm 69
    A Final Note on Differentiation 73

4 The Position of Peasant Agriculture in the Wider Context 78
    Town-Countryside Relations
    as Mediated by Exchange Relations 79
    Town-Countryside Relations as Mediated by Migration 81
    Farming Versus the Processing and Marketing of Food 82

	State-Peasantry Relations	84
	The Balance of Agrarian Growth and Demographic Growth	87
5	Yields	89
	Current Mechanisms of Labour Driven Intensification	95
	The Significance and Reach of Labour Driven Intensification	105
	When Labour Driven Intensification is Blocked	107
	What Propels Labour Driven Intensification?	109
	Intensification and the Role of Agrarian Sciences	110
	Can Peasants Feed the World?	119
6	Repeasantization	125
	Processes and Expressions of Repeasantization	127
	Repeasantization in Western Europe: Resetting the Balances	128
Glossary		132
References		136
Index		149

ICAS Agrarian Change and Peasant Studies Series

The Agrarian Change and Peasant Studies Series by the Initiatives in Critical Agrarian Studies (ICAS) features "state of the art small books on big issues," each of which explain a specific development issue based on key questions. The questions include: What are the current issues and debates in the particular topic? Who are the key scholars/ thinkers and policy practitioners? How have the positions emerged and developed over time? What are the possible future trajectories? What are the key reference materials? Why and how it is important for NGO professionals, social movement activists, official development aid and nongovernmental donor agencies, students, academics, researchers and policy experts to critically engage with the key points explained in the book? Each book combines theoretical and policy-oriented discussion with empirical examples from different national and local settings.

In the book series initiative, the overarching theme, "agrarian change," binds scholars, activists and development practitioners from diverse disciplines and from all parts of the world. "Agrarian change" is meant in its broadest sense, referring to an agrarian-rural-agricultural world that is not de-linked from, but rather taken in the context of, other sectors and geographies: industrial and urban, among others. The focus is on contributing to our understanding of the dynamics of "change"; meaning playing a role not only in (re) interpreting the agrarian world in various ways but also in changing it — with a clear bias for the working classes, for the poor. The agrarian world has been profoundly transformed by the contemporary process of neoliberal globalization, demanding new ways of understanding structural and institutional conditions, as well as new visions of how to change these.

The Initiatives in Critical Agrarian Studies is a worldwide *community* of like-minded scholars, development practitioners and activists who are working on agrarian issues. The ICAS is a *common ground*, a common space for critical scholars, development practitioners and movement activists. It is a pluralist initiative, allowing vibrant exchanges of views from different progressive ideological perspectives. The ICAS responds to the need for an initiative that builds and focuses on *linkages* — between academics, development policy practitioners and social movement activists; between the world's North and South, and South and South; between rural-agricultural and urban-industrial sectors; between experts and non-experts. The ICAS advocates for a *mutually reinforcing* co-production and *mutually beneficial* sharing of knowledge. The ICAS promotes *critical thinking*, which means that conventional assumptions are interrogated, popular propositions critically examined and new ways of questioning composed, proposed and pursued. The ICAS promotes *engaged research and scholarship*; this emphasizes research and scholarship that are both academically interesting and socially relevant, and further, implies taking the side of the poor.

The book series is financially supported by the Inter-Church Organization for Development Cooperation (ICCO), the Netherlands. The series editors are Saturnino M. Borras Jr., Max Spoor and Henry Veltmeyer. Titles in the series are available in multiple languages.

To my grandfathers, Jan Douwe and Fokke,
for teaching me how to delve into things

Acknowledgements

My thanks go to the peasants of Catacaos, Antapampa and Luchadores in Peru; Guarne, Argelia, Sonsón and Chocó in Colombia; Buba and Tombalí in Guinea Bissau; Reggio Emilia, Parma, Campania and Umbria in Italy; Namialo in Mozambique; Mtunzini and Empageni in South Africa; Tras-os-Montes in Portugal; Sangang in China; Londrina and Dois Irmãos in Brazil; and especially to those of Fryslân and the rest of the Netherlands. Individually and collectively, they taught me important lessons. If the following text does not accurately reflect their practices, dreams and explanations, I take sole responsibility.

I also am grateful to Saturnino M. Borras (Jun), who invited me to elaborate this text. I thank Ye Jingzhong for arranging and attending yearly meetings with Jun Borras, Henry Bernstein and many others. These meetings gave me the courage to undertake the writing of this small book on one of the big ideas that needs to be told, re-told and discussed as its relevance has not diminished. I thank Nick Parrott for the careful editing of this text.

Thanks also to the folks at Fernwood Publishing for their efforts in bringing this book out: Errol Sharpe for publishing, Marianne Ward for copyediting, Brenda Conroy for text design, John van der Woude for cover design, Beverley Rach for production coordination and Nancy Malek for promotion. They did an excellent job.

Preface

Peasants and the Art of Farming by Jan Douwe van der Ploeg is the second volume in the Book Series in Agrarian Change by ICAS (Initiatives in Critical Agrarian Studies); the first volume is Henry Bernstein's *Class Dynamics of Agrarian Change*. Indeed Jan Douwe's volume is the perfect follow-up to Henry's book. Together, these two books reaffirm the strategic importance and relevance of agrarian political economy analytical lenses in agrarian studies today. This combined with the very high world-class quality of the books promises that succeeding volumes in the series will be as politically relevant and scientifically rigorous.

A brief explanation about the Book Series in Agrarian Change will help put Jan Douwe's current volume into perspective in relation to the ICAS intellectual and political project.

Today, global poverty remains significantly a rural phenomenon, with three quarters of the world's poor comprised of the rural poor. Thus the problem of global poverty and the challenge of ending poverty, which is a multidimensional issue (economic, political, social, cultural, gender, environmental, and so on), are closely linked to the resistance of working people in the countryside to the system that generates and continues to reproduce the conditions of rural poverty and the struggles of the rural poor for sustainable livelihoods. A concern for and focus on rural development thus remains critical to development thinking. However, this concern and focus does not mean de-linking rural from urban issues. The challenge is to understand better the linkages between them, partly because the pathways out of rural poverty paved by neoliberal policies and the efforts by mainstream international financial and development institutions engaged in and leading the war on global poverty to a large extent simply replace rural with urban forms of poverty.

The mainstream thinking on agrarian studies is generously financed, and so it has been able to dominate the production and publication of research and studies in agrarian issues. Many of the institutions (such as the World Bank) that propagate this thinking have also been able to acquire skills in producing and propagating highly accessible and policy-oriented publications that are widely disseminated worldwide. Critical thinkers in leading academic institutions are able to and do challenge this mainstream current in many ways but are generally confined to the academic circles with limited popular reach and impact.

There remains a significant gap in addressing the need of academics (teachers, scholars and students), social movement activists and development practitioners in the global South and the North for scientifically rigorous yet accessible, politically relevant, policy-oriented and affordable books in critical agrarian studies. In response to this need ICAS is launching this series of "state of the art" small books, books that will explain a specific development issue based on key questions, including: what are the current issues and debates in this particular topic; who are the key scholars/thinkers and actual policy practitioners; how have such positions emerged and developed over time; what are the possible future trajectories; what are the key reference materials; and why and how it is important for NGO professionals, social movement activities, official development aid circle and nongovernmental donor agencies, students, academics, researchers and policy experts to critically engage with the key points explained in the book. Each book will combine theoretical and policy-oriented discussion with empirical examples from different national and local settings.

The Book Series in Agrarian Change will be available in multiple languages, at least initially in three languages in addition to English, namely, Chinese, Spanish and Portuguese. The Chinese edition is in partnership with the College of Humanities and Development of the China Agricultural University in Beijing coordinated by Ye Jingzhong, the Spanish edition is coordinated by the Ph.D. program in Development Studies at the Autonomous University of Zacatecas in Mexico coordinated by Raúl Delgado Wise, and the Portuguese edition with the Universidade Estadual Paulista, Presidente Prudente (UNESP) in Brazil coordinated by Bernardo Mançano Fernandes.

Given this explanation of the context for and objectives of the Book Series, one can easily understand why we are very pleased and honoured to have Henry Bernstein's and Jan Douwe van der Ploeg's books as the first and second in the series, respectively: together they are a perfect fit in terms of theme, accessibility, relevance and rigour. We are excited and optimistic about the bright future of the Book Series!

Saturnino M. Borras Jr., Max Spoor and Henry Veltmeyer
ICAS Book Series Editors

1

Peasants and Social Transformations

A Divisive Issue

When it comes to the peasant question, the radical left has been deeply divided. In several ways it still is — although there definitely are indications in political and scientific debates, in new social movements and in socio-material reality itself, that the great divide is increasingly being bridged. And if this sounds too optimistic, then we may probably argue that the divide is not so much being bridged but increasingly becoming less relevant (which may also represent a way to resolve controversies, especially political ones). The earlier controversies are fading away because we are witnessing, in many places around the world, new developmental tendencies that definitely go beyond the limits of previous debates.

Historically, the main controversies have been strongly associated with two leading spokesmen, Vladimir Lenin and Alexander Chayanov who, in the first decades of the twentieth century engaged in sharp polemics that reflected different interests and prospects that already had lain dormant in Russian society for a long time and that drastically came to the fore in the aftermath of the 1917 revolution. At that time Russia was basically an agrarian nation. Industry made up just a small part of the national economy. Peasants hugely outnumbered industrial workers and, although capitalist farm enterprises were emerging (and their significance was hotly debated), peasants made up the large majority of rural dwellers. Peasant communities provided the framework that regulated everyday life for the majority of Russians. Lenin (and more generally, the bolshevists) and Chayanov (representing, in a way, the *narodniki*[1]) interpreted this reality in different ways, taking different positions about the role of different social groups (particularly the peasantry), which created fierce controversies about the future of Russian society.

Originally, the great divide centred on several, strongly interrelated issues. The most important ones concerned, in the first place, the definition of the class position of the peasantry — a question that clearly related to practical issues, such as the nature of coalitions and the role different parts of the population might play in revolutionary processes. Second, there was much debate on the stability of peasant-like forms (or "modes") of production (see also Bernstein 2009). Would they inevitably disintegrate, or would it be possible for them to be reproduced over time? Or would there be unequal but combined processes of disappearance and reconstitution? Third, should those engaged in the transition toward socialism regard peasant agriculture as something to be continued or transformed? Are peasant modes of production a promising way to produce food and make significant and substantial contributions to the development of society as a whole? Or are other forms of production, such as large state-controlled cooperatives (be it kolkhozes, peoples' communes or whatever) far superior? Is the peasantry a hindrance to change, insofar as it will struggle to block the transition to such supposedly superior forms? Or might it become a main driver of the transformations needed in the countryside?

Today, that is, in the beginning of the twenty-first century, many of these questions might seem terribly outdated, especially when they are exclusively linked to the Russian situation of the post 1917 period. Yet, we have to take into account that

(a) The controversy was in no way limited to Russia. The main spokesmen of that time also referred to, and tried to integrate into their analyses, different experiences from other places: America, Germany (notably Prussia), Switzerland, Czechoslovakia, Italy and the Low Countries. Equally, the debate quickly extended to a global one that ranged from East to West and from North to South. Wherever power was seized or major regime shifts occurred, the question was asked whether socialism (or more generally, a better society) could be constructed by giving peasants a prominent role in the overall process of rural development. This question arose with insistence, especially in those places where peasants had been in the forefront of revolutionary struggles, from Mexico, to China, Cuba and Vietnam (Wolf 1969). In these countries the debates often came down to another important question: how should land reform be organized? These were far from just theoretical questions. They were of immediate concern in Mexico in the 1930s and then in Italy in the immediate postwar years when a land reform was designed and partly implemented. In 1974 it was a central concern in Portugal and soon after in Angola, Mozambique and Guinea-Bissau, in Cuba after Castro's revolution and then again in the early 2010s and in China in the second half of the 1940s and then again from 1978 onwards. The same debate emerged in Vietnam in 1954 and 1986, the year of Doi Moi. In Japan the debate started after World War II and never disappeared from the agenda. In the Philippines it was a major issue in the 1950s, was triggered again by the 1986 elections and intensified during and after the Aquino reform of 1988. Latin America witnessed similar debates and, although there were specific foci time and again (such as the period of the Legas Camponesas in Brazil and the radical Reforma Agraria in Peru), in the end the debate covered the whole of the continent and helped to shape its agricultural sectors of today. The many land reforms that swept the continent can be seen as a struggle between the *campesinistas* (who took Chayanovian stances) and the *descampesinistas* (who took Leninist positions) and vice versa. Thus, the controversy that arose first in Russia in 1917 was repeated time and again. In the words of Kerblay (1966: xxxvi): "While Lenin ... demanded prompt confiscation of the large estates ... and nationalization of the land, including that of the peasants, the League for Agrarian Reform [Chayanov was a member of its executive committee] was content to propose the transfer of all land to peasant farms."

The same debate reappeared, albeit in slightly different terms, when it came to the (potential) role of peasant communities. The *mir*, the Russian peasant communities, had been an important point of reference for radical political movements in Russia. Elsewhere, the potential role of such communities in processes of transition was also acknowledged. For instance Mariátegui, a leading Latin American radical thinker, argued: "The peasant community embodies an effective capacity for development and transformation" (1928: 87).

(b) The controversies did not remain limited solely to agrarian issues but also extended to many new questions. For example, in

Peru this was "*el problema del indio*," the question of the indigenous Quecha and Aymara speaking population that keep livestock in the Andean mountains and who are badly discriminated against, exploited and oppressed. Mariátegui skilfully related this "question of the Indians" to the agrarian question, arguing that the multidimensional neglect and subordination of the indigenous population could only be resolved through a radical change in the social relations of production in the countryside. The same occurred, for example, in Italy, where Gramsci tied the "southern question" (in the south of Italy, large landholdings exerted a stranglehold effect that increasingly became a burden for the whole of Italy) to the "agrarian question." The more so since the 1920 Turin uprising had made clear that as long as "the workers stood alone, they were indeed automatically defeated unless they could link their forces with those of the surrounding countryside, to which they were connected in any case by multiple family ties" (Lawner 1975: 28).[2] Far later a similar extension of the peasant question was formulated in China: the *san nong* (three rural issues) policy linked the peasant issue to total agricultural production and the attractiveness of village life (Ye et al. 2010).

The debate on the peasantry also extended toward debates on the contribution of agriculture to the development of society as a whole.[3] Agriculture could be heavily squeezed in order to feed capital accumulation in urban industry and to provide the required cheap labour. But some outlined other alternatives. A prosperous countryside (as opposed to a squeezed agriculture) could very well become an attractive internal market and thus offer strong support to industrialization (Kay 2009). Another debate, that emerged far later, was on sustainability. It is interesting that the first ones who initiated this debate were clearly located in the Chayanovian tradition, such as Vries (1948), for example. Today any discussion on the path to sustainability necessarily has to debate the role of the peasantries. Yet another debate that constantly reappears is the one on poverty (see, for example, IFAD 2010). Tragically, the number of poor people in the world continues to increase steadily, reaching an estimated 1.4 billion in 2010. Typically, 70 percent of the poor of the world are rural; they live in the countryside and depend, more or less, on agricultural activities. Food scarcities are a frequent and recurrent phenomenon, and it is expected that world food production needs to be doubled by 2050 when world population is assumed to peak. However, neither the short term food shortages nor the long term need for agricultural growth are translated into opportunities for these rural poor. Instead, they trigger new corporate investments (land grabbing being the most visible expression) that further damage and undermine the livelihoods of many rural people.

(c) Last but not least, it became increasingly clear that the initial questions and the extended fields of debate that were later added were not only relevant for the radical left. Other political currents, including institutionalized science, had to face and deal with the same issues. All these domains have become divided over exactly the same issues, and none has been capable of resolving the associated controversies. Badly equipped insofar as the main concepts and fields of interests are concerned and ignoring the potentially powerful contributions from Chayanov, scientific disciplines as diverse as agrarian economics, development economics, rural sociology and peasant studies as well as institutions such as the World Bank and the UN Food and Agricultural Organization (FAO) have not been able to contribute much to resolving these issues (Shanin 1986, 2009). The specific solution arrived at by some, i.e., declaring the death of the peasantry, also did not turn out to be very helpful.

This book does not aim to be an extensive reconstruction of the historical polemics, nor does it pretend to resolve them in an *ex post* way. My aim is to synthesize the core of the Chayanovian approach and link it to current issues that are central to many new, rural movements.

Central to the Chayanovian approach is the observation that although the peasant unit of production is conditioned and affected by the capitalist context in which it is operating, it is not directly governed by it. Instead, it is governed through a set of balances. These balances link the peasant unit, its operation and its development to the wider capitalist context but in complex and definitively distinctive ways. These balances are ordering principles. They shape and reshape the way fields are worked, cattle are bred, irrigation works are constructed and how identities and mutual relations unfold and materialize. The range and complexity of balances involved, which are continuously reassessed, gives rise to the impressive heterogeneity of peasant agriculture and creates a permanent ambiguity. On one hand the peasant is downtrodden and misunderstood, on the other he or she is indispensable and proud. The peasantry both suffers and resists: sometimes at different moments, sometimes simultaneously. Similar confusion and apparent contradictions apply to agriculture as a whole; it sometimes witnesses processes and periods of depeasantization and sometimes of repeasantization. All this can be traced back to the complex interactions between different balances and how each balance is cast and recast by different actors (peasants, their families, communities, interest groups, traders, banks, state apparatuses, agro-industries, etc.)

Chayanov focused on two balances (one of labour and consumption, the other of drudgery and utility) that are to be equilibrated within each peasant farm in a way that is singular to that farm and to the needs and prospects of the peasant family living and working there. These balances combine incommensurable entities (e.g., labour and consumption) that are necessarily related to each other. Consequently, the balances constitute "*mutual* relationships" (Chayanov 1966: 102, italics added). Building on this approach I will discuss a far wider array of balances — some internal to today's peasant farms, others more general insofar as they link peasant agriculture with the dynamics taking place in the wider surroundings. In doing so I am extending Chayanov's approach. That is to say, I seek to go beyond the many time and space bounded limitations that are inherent to Chayanov's work (and of which he was well aware)[4] and identify the balances that operate as the main ordering principles in today's peasant agriculture. I will also try to indicate how peasant agriculture can contribute to responding to some of the big challenges humankind is facing; such responses depend very much on an adequate coordination of different balances — at least, if sufficient "space" (Halamska 2004) is granted to, or conquered by, the different peasantries of this world.

In the rich tradition of peasant studies that evolved worldwide during the twentieth century, many balances have been identified. I will show that the art of farming,[5] an expression literally used by Chayanov in his *Social Agronomy* (1924: 6), comes down to the skilful coordination and intertwinement of the interacting balances (see, for example, Chayanov 1966: 80, 81, 198, 203). Through this coordination peasant farms are turned into a "well working whole" as Dirk Roep (2000) argued in relation to Dutch peasant farms operating at the turning of the millennium.[6] I will also try to demonstrate that the assessed equilibria are far from static. They are dynamic: they translate the emancipation aspirations of the peasantry into ongoing agrarian and rural development — unless such development is blocked by other relations and circumstances. And finally I will demonstrate that the coordination and intertwinement of the different balances does not separate the peasant farm from its politico-economic environment. Instead, it links them to, and simultaneously distantiates them from, this environment. Every balance is a unity of initially incommensurable entities that nonetheless need to be combined and aligned. Thus, there is the need to find the best possible equilibrium. This implies trade-offs and often generates frictions. Operating a balance and trying to reassess it (if needed) often translates into, or can fuel, social struggle. This is true especially when we take into account the different forms of social struggle.

Together the different balances constitute a complex system of thought that

> relies on two basic principles: dualism and relativism. Dualism is a way of perceiving opposites that can be divided but, at the same time, remain complementary. For example, all the territories in the Andes are divided in high and low, with soils that are principally cold and warm. But if one applies the principle of relativism these opposites lose their absolute delimitation. For example, high terrain becomes low when the point of reference and perception of the peasant is on the former — for an external observer a clear sign of logical inconsistence, but for the peasant a smooth passage to blend opposite values. The point of reference is the middle. (Salas and Tilmann 1990: 9–10)

The art of farming greatly depends on using good judgement to assess the different balances. "We can affirm that the art of farming is rooted in the most appropriate use of the many particularities that are entailed in his farm" (Chayanov 1924: 6). These particularities are understood and managed as part of a balance; together they flow into an equilibrium that links particularities, for example the available land, the number of cattle, the number of people able to help in the labour process, the savings and investments, etc., into one well working whole. A balance is a regulatory device (a bit like a thermostat). It continuously registers relevant information (e.g., the temperature of the room) and translates this into appropriate responses and reactions (e.g., increase, decrease, postpone or completely stop the heating). Significantly, in his discussion of these balances Chayanov first and foremost takes into account the features (and more generally the interests, prospects and experiences) of the peasant family. When we talk about the balance of labour and consumption, we are not talking about abstract consumption, but about the specific (or concrete) consumption needs of a particular family. The same applies to labour: it is the amount and quality of labour that a particular peasant family (located in a particular situation) is able and willing to deliver. And finally the family is a specific constellation, characterized by specific features, such as the consumer/worker ratio (which will be explained further on). But it is the peasant him or herself who adjusts and readjusts the different balances.

Thus we can further extend the analogy of the thermostat to illustrate the specificity of the Chayanovian balances. First, whilst a thermostat is fed with and reacts to objective data (e.g., the temperature of the room in degrees Celsius) — that are non-negotiable and not open to any subjective evaluation whatsoever — Chayanovian balances critically take into account the way particular features are perceived by the involved actors themselves (i.e., how the temperature in the room is experienced by those present in it). This is far more complicated than working with just objective data. Second, whilst the thermostat is a completely automated device that can operate without the permanent presence or intervention of any actor, the Chayanovian type of balance is critically operated by an actor (or group of actors) — i.e., by a craftsman who understands farming. Third, the thermostat applies the inbuilt algorithm in a linear, unequivocal and non-negotiable way. The thermostat cannot produce diversity. Eighteen degrees Celsius is exactly the same on Monday morning as it is on Wednesday evening. But when assessing a Chayanovian balance, the involved actors usually operate rules that are part of the cultural repertoire of their community or professional group. Such rules always imply an active interpretation and adequate application to specific situations. They are not applied in a mechanical, one-to-one relation. There is no simple mathematics in peasant farming. This is one of the reasons why diversity emerges. It also explains why farmers often quarrel.

In synthesis: Chayanovian balances critically take into account the specific situation of the singular peasant family and peasant farm. As such these balances are actor-dependent rather than automated devices. The operation of a balance (that is, its application to a singular situation in order to generate a solution) involves actors being able to interpret rules and situations and to make the appropriate decisions. This raises the critical question of gender relations, although these are not taken into account in the original work of Chayanov. However, since the 1980s, a lot of path-breaking work has been done in this respect (see, for example, Rooij 1994; Agarwal 1997). Another set of internal familial social relations that will be increasingly decisive for the future of farming concerns intergenerational renewal and particularly the prospects for youth in agriculture. Here much work still needs to be done (White 2011; Savarese 2012).

Most of the balances discussed in this small book regard relations (be they direct or indirect) between the peasant unit and the wider environment. The latter often affects the peasant unit in adverse ways. This makes the regulation of the relevant balances a delicate affair. For it is not only the peasant family that is searching for the best possible equilibrium. External agencies (such as agro-industries, banks, trading companies, retail chains, technicians and extensionists) are also actively intervening, trying to reassess the different balances in ways that better correspond with their own rationale, even if this is detrimental for the direct producers. Thus, many of the balances to be discussed here are the result of, or represent, antagonisms. They are the arenas where the representatives of different sets of interests meet, struggle, align and/or negotiate. Assessing a precise equilibrium for each and every of the many interlinked trade-offs (or in

Chayanovian terms, balances) thus becomes part of wider struggles. The discussion of the different balances equally makes clear that peasants' struggles are not restricted to the streets, to occupying central squares in the capitals or setting fire to a McDonald's — they are also, equally, struggling when trying to improve a field or to construct a communal irrigation system.

Chayanovian balances are what constitute and regulate farming. They shape and reshape, within particular time and place bounded contexts, the layout and the fertility of fields, the number and type of cattle, the yields rendered by crops and animals, etc. In short "the organizational plan of the peasant farm" (1966: 118) and its unfolding over time are regulated by and through the different balances. If beautiful fields, "well-bred" manure, good grain harvests and heifers that provide good offspring are all expressions of the art of farming, then mastering, fine-tuning and creatively combining the different balances form the core of this art.[7] They are the instruments used by the artist in order to make his masterpiece.

But this doesn't just occur on the farm. Peasant families employ the different balances to translate their interests, prospects and aspirations into a script that also specifies the way the farm is to be developed in the future, the way to operate in the market places, in village meetings, etc.

Peasants often select equilibria that serve to distantiate the organization, operation and development of the peasant farm from the immediacies of the market, thus protecting (albeit only partially) the productive unit, the peasant family and the community to which they belong, from the many threats within these markets. Thus, the balances that translate into specific equilibria might be understood as a kind of Polanyi type of "anti-market device": they help peasants and peasant agriculture to swing away from the markets whenever and wherever this is needed. Thus, it is not only the state that intervenes to correct any major misbalances that occur between economy, ecology and society. It is a particular part of civil society (i.e., the peasantry) that "intervenes" in the development of agriculture, pulling it away from a route determined by the economy only. It does so through mastering and tuning the different balances. The peasantry's active control over the different balances makes agriculture into a constellation that is more productive, provides more employment and offers many people more autonomy and room for self-management than would be the case if farming were controlled solely by markets and/ or capital-labour relations.

The Political Relevance of Peasant Theory

The historical debates on peasants and peasant agriculture cannot be set aside as irrelevant or outdated quarrels. They reflect and relate to different pathways to construct and develop specific sociomaterial realities. The basic dilemmas are still present in today's world — maybe more than ever (see, for example, Mazoyer and Roudart [2006] who argue that the general economic crisis of today's capitalism cannot be solved without an adequate response to the massive poverty to which large parts of the rural population are condemned). The same applies to the core of the work of Chayanov. Thirty years ago Paul Durrenberger asked "why [we] should attend to his work more than 50 years later?" (1984: 1). His answer to this question still seems valid: "The simplest answer is that Chayanov developed an analysis of peasant farm economics and household production units that is relevant wherever and whenever we find such forms" (ibid.).

Reconsidering the "art of farming" more than one hundred years after the first debates divided the radical left of that time is important, I think, for at least five reasons.

First, there is an epistemological reason. As Mottura (1988: 7) exposes in an intelligent introduction to Chayanov, there are basically two positions toward the peasantry, now as well as in the past. One is an uncritical belief (like the populist position in the past and the "choosing the side of the peasants" position today), the other is outright aversion. Between the two there is no critical position, let alone a critical theory. As I tried to argue in *The New Peasantries* (2008), peasant agriculture is a practice without a theory. Hegemonic thinking is arrogant toward and ignorant about peasantries and peasant modes of farming. The modern world relates to peasant realities through either belief or aversion. This makes those realities into uneasy phenomena, awkward realities indeed. Chayanov is the exception in this panorama. He holds the promise that we might develop an understanding of the peasantry and even possibly construct a viable critical theory. Chayanov's relationship with the Russian peasantry can be characterized in several keywords. Curiosity is the first and foremost. Empirical curiosity: what drives these people? What are the potentials entailed in their ways of farming? How do they relate to each other? What can they contribute to society?[8] It is telling that Chayanov tries to find the answers within the peasantry — peasants and peasant agriculture are not externally determined and governed by "general laws." Hence, an empirical inquiry into the dynamics of the peasantry is crucially needed for the elaboration of an adequate theory. This comes with other key elements: academic rigour, involvement and hope.

Curiosity feeding into well-grounded empirical research has been the vehicle, in the many decades that followed, for a nearly relentless reinvention of the Chayanovian position. Many researchers and intellectuals tightly linked to the peasantry discovered only afterwards the value and strength of the original work of Chayanov, thus contributing to what we now refer to as the Chayanovian approach.

Second, today's world is witnessing massive, albeit highly varied, processes of repeasantization. There are notable expressions of this in the "return" to small family farms in China, Vietnam and other southeast Asian countries — a landslide that has seen more than 250 million peasant farms reappear and that turned China into "an academic goldmine" for peasant studies (Deng 2009: 13). Another remarkable process took place in Brazil, where the rural exodus (that started under the military dictatorship of the 1970s) was reversed through a massive movement of hundreds of thousands of poor people, mainly but not only from the miserable and dangerous *favelas*, toward the countryside. They occupied large tracts of land that were finally converted, after lengthy and tough fights, into many new peasant units. According to the last two national censuses (1995–1996 and 2006), the number of small holdings grew by some 400,000 (representing an increase of 10 percent in the total number of farms [MDA 2009]). Together, these newly created peasant farms cover an area of 32 million hectares, "which equals the total agricultural area of Switzerland, Portugal, Belgium, Denmark and the Netherlands taken together" (Cassel 2007). Other expressions of repeasantization can be found in Europe. I will describe these in more detail in chapter 6.

Third, there is the rise of new, proud and powerful movements that operate internationally and consequently are often referred to as "transnational agrarian movements" or TAMs (Borras et al. 2008) such as Via Campesina, which literally means "Peasant Path." Their growth has coincided with (and doubtless provoked) increasing attention to the peasant issue from established NGOs as well as international organizations that operate within the UN framework. *Les paysans son de retour* (the peasants are back) is the title of a 2005 book by Perez-Vitoria. Indeed they are back, both in practice and in policy.

In the fourth place, there is the growing insight that peasant agriculture holds an important response to many of the new scarcities (food, water, energy, productive employment, etc.) that are threatening the future of our planet (I will come back to this response in chapter 5). Peasant agriculture may also have a role to play in helping to mitigate climate change, since as Via Campesina claims, peasant agriculture has a "cooling" instead of a heating effect. The same applies when thinking about the economic and financial crises, which contribute considerably to volatility in the markets: here peasant agriculture comes to the fore as it provides a strongly resilient form of food production.

In the fifth and final place we have to take into account that, over the past decades, radical theory has moved beyond many categories that were intimately linked to the genesis and heydays of industrial capitalism. The once classical proletariat dissipated into multiple "classes of labour" (Bernstein 2010a); the classical factory is no longer the central location of the confrontation between labour and capital. The antagonism between the two now pops up in many, widely distributed places and takes new and often intriguing forms (Hardt and Negri 2004). Political theories that try to seriously describe these changes (e.g., Harvey 2010 and Holloway 2002 and 2010) have developed new approaches that shed new light on old issues, sometimes offering unexpected perspectives.

These newly emerging approaches do not only highlight, albeit indirectly, the relevance of the initial work of Chayanov, they also allow for its further elaboration. By combining Chayanov and much of the subsequent Chayanovian work with these new political approaches, we can improve our understanding of the many rural struggles occurring in the world today as new rural movements try to change the world.

By way of introduction I will briefly refer here to three concepts (I will return to them in the final chapter). The first is multitude. The peasantries of today's world are multitudes. They master the art of not being governed (Scott 2009; see also Mendras 1987); they are highly heterogeneous; the sources that inspire the ordering of their labour processes extend far beyond the logic of the market: nature, society and cultural repertoires are all equally important ordering principles (as I will discuss throughout this book). They resist the separation of the process of production into separate tasks, just as they redress the tendency to externalize many such tasks. They create commons — a second important concept.[9] Commons — such as occupied land in Brazil, shared seed reservoirs throughout Latin America and Africa, irrigation works in China, new town-countryside relations in Europe and newly constructed nested markets all over the world — turn out to be highly productive and offer a potentially convincing alternative to corporate capital. Third, there is the concept of interstices, i.e., the places where antagonisms occur. Interstices are the cracks in the global system, the structural holes that emerge as a result of massive processes of exclusion. They are the voids that the state apparatuses cannot regulate through their institutional machinery. Some of these interstices just emerge, others are actively created from the often chaotic and contradictory realities in which we are all moving.

Peasant families operate at the intersection of several interstices. The first is, of course, represented by the fact that their labour is not wage labour. It is not directly subordinated to capital even though capital does try to construct and effectuate complex and often deeply penetrating mechanisms to control peasant labour. Through the active and knowledgeable adjustment of the many balances that underlie today's peasant farm, many peasants distantiate the operation and development of their farms from the "logic of capital." That is, they create interstices. And they increasingly interlink with others creating and operating within other interstices — often giving birth to new social movements. More generally, interstices are places of permanent struggle, they are cradles of resistance and sometimes emerge as places where solid alternatives to capitalist arrangements are forged. They are the places where the multitudes are located and where singularity is produced and reproduced. I will come back to these issues in the final chapter.

Peasant Agriculture and Capitalism

Chayanov (1966: 222) made it abundantly clear that the peasant farm "exists within an economy dominated by capitalist relations; it is drawn into commodity production and is a petty commodity producer, selling and buying at prices laid down by commodity capitalism and its circulating capital might be based on bank loans." "Through these connections, every small peasant undertaking becomes an organic part of the world economy, experiences the effects of the world's general economic life, is powerfully directed in its organization by the capitalist world's economic demands, and, in turn, together with millions like it, affects the whole system of the world economy" (ibid.: 258).

In brief, peasant farms are part of the capitalist system. However, it is also true that a peasant farm (a) is a subordinated part (see, for example, ibid.: 257); (b) in itself it is not a capitalist unit of production; and (c) it operates in a way that often is distinctively different from the way in which capitalist farm enterprises are managed.

The peasant farm is not structured as a capitalist enterprise; it is not grounded on a capital-labour relation. Labour, within the peasant farm, is not wage labour. And capital is not capital in the Marxist sense (i.e., it is not capital that needs to produce surplus value to be invested in order to produce more surplus value). In the peasant farm the "capital" is the available tools, buildings, animals and savings. But this "capital" is definitely not "a value that produces surplus value," which is how Kautsky (1974: 65) understood it. The buildings, equipment, etc., are instruments (or means) to facilitate and to improve the labour process (see also box 5.1). It is the absence of the capital-labour relation that turns particular units of agricultural production into peasant farms. This is the decisive defining factor in the Chayanovian approach.

The specific internal structure of a peasant farm means it is often operated in ways that differ decisively from capitalist farm enterprises — and it is precisely this difference that is of great importance. In the words of Chayanov (1966: 89), "The peasant farm continues to produce where capitalist farms stop." Thorner (1966: xviii) says, "In conditions where capitalist farms would go bankrupt, peasant families could work longer hours, sell at lower prices, obtain no net surplus, and yet manage to carry on with their farming, year after year. For these reasons Chayanov concluded that the competitive power of the peasant family versus large-scale capitalist farms was much greater than had been foreseen in the writings of Marx, Kautsky, Lenin and their successors." Mariátegui (1928: 103) reinforces this point: "We see everywhere around us that the large land owner is not interested in the physical productivity of the land but only in its profitability."

Peasant agriculture is part of capitalism. But it is an uneasy part. It generates interstices and frictions. It is the cradle of resistance that produces alternatives that act as a permanent critique of dominant patterns. It goes where capitalist farms cannot go. Peasant agriculture is "anaerobic" (Paz 2006); it can survive without the oxygen of profit that corporate agriculture so badly needs. Being part of capitalism also makes uneasy farms. Through the balances several of the main contradictions penetrate into the peasant farm. Consequently there are struggles within the peasant family as well, just as in the peasantry as a whole.

All this implies that it is not only possible (as convincingly argued by Little [1989]) to combine politico-economic analysis (to research the context and how it translates into the peasant farm) and the Chayanovian approach (for understanding the specific translation and the development of responses), it often is necessary to do so. The aim is not to detect all manner of hairsplitting differences and supposed incompatibilities between the two but rather to forge them into one strong theoretical tool.

This book rejects the (dominant) view of the peasantry as a phenomenon that is necessarily limited to the past and to the periphery. Nor does it accept the view that the modernization of agriculture in the West has eliminated peasant-like ways of farming. True, peasant societies have disappeared, just as a new way of farming has emerged based on the entrepreneurial model (a model that involves a complete reshuffling of many of the main balances). But the peasant way of farming has continued, attuning itself to new circumstances, and since the early 1990s it has been revitalized, strengthened and extended; in short, it has experienced a renaissance. Many farmers (I use the term farmer as a generic concept that embraces many different types) all over the world continued or started again to produce as peasants. They do so in many different ways that correspond to the exigencies, difficulties and possibilities facing them in the early twenty-first century.

The peasantries of, say, Latin America and northwest Europe are indeed very different entities, and any attempt to group them in one single analytical category — "peasants" — rightly raises the question "what do they have in common"? Bernstein (2010a: 112) examined this question by asking, "Is there any common social relation with capital?" I think the argument that peasants share certain common conditions of existence vis-à-vis corporate capital and therefore have a common basis for collective action in the pursuit of common interests provides a solid basis for legitimately grouping them together as a single entity (see also Bernstein 2010b: 308).

What Makes Chayanov a "Genius"?

I will refrain from providing a biography of Chayanov. Others have done so (Kerblay 1966; Sperotto 1988; Sevilla Guzman 1990; Danilov 1991; Abramovay 1998; Shanin 2009; Wanderley 2009) and did a much better job than I could ever hope to do. But I do want to stress his genius was not divinely inspired: he was, as everybody (and maybe especially the geniuses among us), a product of his circumstances. First, there was the specific historical background that included the endless but highly diverse Russian countryside, the economic depression in the mid nineteenth century, the many *mirs* (peasant communities) and radical political movements (mostly known under the umbrella of *narodniki*) that envisaged a Russian future that would be built on the peasantry and be constructed together with them (Sevilla Guzman and González de Molina [2005] provide a succinct overview of these movements and their programs). Chayanov was more than familiar with this background. He also knew peasant life through many daily encounters, as is evident from many fragments in *Social Agronomy*, a work that is only available in German and thus hardly known elsewhere. But he also had another way of knowing about peasant agriculture and its dynamics: a way that was relatively unique at that time.

Second, Chayanov had access to a unique database, the *zemstov* statistics. Auhagen, who wrote the preface to the first German translation of *The Theory of Peasant Agriculture*, noted, "I don't know of any country that has such a rich agricultural database as Russia" (Auhagen 1923: 1). And Chayanov noted, I guess proudly, that Karl Marx himself expressed his admiration for and interest in these *zemstov* statistics (1923: 7). This rich data allowed for the exploration and analysis of empirical patterns that reflected the operation of different balances. Together with well-developed methods of statistical analysis, the availability of this rich material created a unique opportunity.

Third, Chayanov had the advantage of working and living in a transitional period that started with the Bolshevik Revolution of 1917 — although this same advantage turned out in the end to have deadly consequences. He was arrested, suffered a show trial and died in the Gulag archipelago. However, before such tragic events became a systematic feature of Soviet society, post revolutionary Russia was a lively cauldron of fermenting ideas, where the prospect of far-reaching rural change was extensively discussed. Chayanov, who was involved at many levels, was one of those who embodied the optimism of these movements.

Together these three ingredients provided a unique mix of circumstances that were translated by Chayanov into at least three major and, at that time, absolutely novel lines of reasoning:

1) A theory of peasant agriculture that included a first attempt to unravel the dynamics of the individual peasant farm and of peasant agriculture as a whole. This microlevel theory was combined with a more general discussion (at the macro level) in which the "isolated state" (or the "island") was used as a metaphor, with a strong hint of the importance of carefully regulating the internal (or national) market, especially when it comes to international

trade. Chayanov also developed a utopian view about how peasant agriculture could possibly unfold within a prosperous society located somewhere in the future. He did this anonymously, using the pseudonym "Ivan Kremnev," in a 1920 novel that describes the journey of "Brother Alexis" (Chayanov 1976).

2) An outline of what he called "social agronomy," which several authors claim to be the starting point of rural extension and of extension studies. It is also an outline of an agronomy that recognizes the centrality of the interactions between, and mutual transformation of, people and living nature (rather than viewing agriculture as governed solely by "laws of nature").

3) A theory of vertical cooperation (as opposed to the "horizontal cooperation" imposed by the "collectivization" that followed later), which is an early example of a theory of transition (Kerblay 1985).

The last of these lines of reasoning, vertical cooperation, deserves a fuller explanation. It refers to the building of strong cooperatives on both the upstream and the downstream sides of the peasant farm. On the upstream side these might be cooperatives that produce and deliver inputs (e.g., fertilizers, machines, credit facilities) to peasant farms. On the downstream side they would process and commercialize the different produce from peasant farms. Such "cooperatives render to small enterprises all the benefits of the large ones" (Chayanov 1988: 155). In the years preceding the 1917 revolution, the cooperative movement had gained considerable momentum in the Russian countryside. Building on this extensive web of cooperatives was a cornerstone for a far wider political project: the transition of Russia, a project that was anticipated to involve radical agrarian reform. This transitional project was to be guided by three clear objectives: 1) increase agricultural production as much as possible, thus contributing to the overall growth of the national economy;[10] 2) strive to maximize the productivity of agricultural labour; and 3) distribute national income more equitably. In Chayanov's view this transition critically needed to build on the peasantry[11] and to be driven forward by the peasantry itself: "Before us there are millions of peasants, with their own habits, their own ideas about farming.

These are men that nobody can command. They do whatever they do according to their own willingness and according to their own concepts" (ibid.). In this and other respects Chayanov came close to the peasant based political project that Karl Marx proposed in a letter dated March 8, 1881 (Marx and Engels 1975: 346). In this letter Marx pointed out that there is no universal theory of historical development. The Russian peasants' communes, he argued, had the capacity to proceed directly toward communism.[12]

This view was a considerable step away from his earlier thinking. In *The Eighteenth Brumaire*, Marx (1963: 124) argued that

> insofar as there is merely a local interconnection among ... small-holding peasants, and the identity of their interests begets no community, no national bond and no political organization among them, they do not form a class. They are consequently incapable of enforcing their class interest in their own name ... They cannot represent themselves, they must be represented.

Building on this we can now argue that once peasants communicate (which is now abundantly the case) and share a joint political project intended to transform the countryside, they constitute themselves into a class — one that might be very capable of putting its imprint on the transitions happening at the time. And this is what is currently happening within and because of the new transnational peasant movements (such as Via Campesina) and their radical agendas for change.

A Proof of Pedigree

Many scientists have explicitly built their work on that of Alexander Vasil'evich Chayanov. And even more have, without knowing his work, "reinvented" the Chayanovian approach, basically because thorough empirical research often induces conceptual frameworks that carry remarkable similarities to Chayanov's theoretical stance. In figure 1.1 I have tried to assemble the best-known scholars that have drawn strongly, albeit often critically, on the work of Chayanov. The pedigree provided is far from complete but serves to illustrate Chayanov's enduring influence. I mainly present it to provide a helping hand to

1 / PEASANTS AND SOCIAL TRANSFORMATIONS

Figure 1.1 A Graphical Sketch of the Chayanovian Tradition

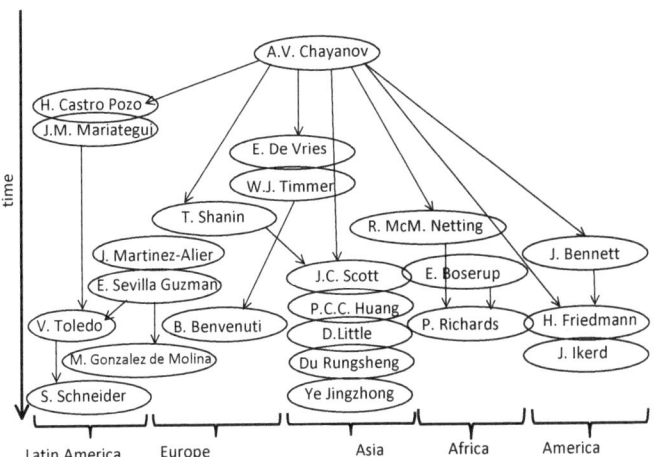

young scientists and social activists who have just started digging into peasant studies. The geographical specification does not refer to place of birth or residence, but to the main place(s) where these scholars did their empirical fieldwork. They are nearly all referred to or quoted in this book. Some of them (such as Martinez-Alier, Sevilla Guzman, Vries and Netting) have worked on more than one continent. The time period roughly extends from 1900 to the present.

Notes

1. This was a Russian revolutionary movement from the late nineteenth and early twentieth centuries. It aimed for an egalitarian society, strongly rooted in Russian peasant communities. At the beginning of the twentieth century the ideas of this movement were articulated by the Social Revolutionary Party, which had strong support in the Russian countryside (see also Martinez-Alier 1991).
2. "Lenin's theory had been that since the peasants made up the majority of the population, it was essential to secure their support, or at least their neutrality. But in Italy, it was clear that the working class would be in a position to realize its vision of the state and democracy only if it assumed the burden of the gravest problem in national history … : the southern question" (Lawner 1975: 28).

3. This debate is historically known as the Preobrazensky-Bucharin debate. It later returned in many guises. A current expression, which clearly follows the same line, can be found in Jackson 2009.
4. Thorner notes that, in this respect, "Chayanov himself conceded that his theory worked better for thinly populated countries than for densely populated ones. It also worked better in countries where the agrarian structure had been shaken up ... than in countries with a more rigid agrarian structure. Where the peasants could not readily buy or take in more land, his theory would have to be seriously modified" (1966: xxi). There were several other limitations. Wherever needed I will refer to these in the next chapters.
5. *The Art of Farming* by Columella (republished in 1977), a Spanish agronomist, is the oldest Western agronomic handbook and is very well written.
6. It is interesting to note that, nearly one hundred years before, Chayanov used a similar image when referring to the peasant farm as a "machine" (1966: 44). When writing his book, Roep was not aware of this. But being a son of a peasant family he was, through everyday life experience, very well acquainted with this aspect of farming.
7. Mastering the different balances is a core element in cultural repertoires of peasant societies. Many equilibria are condensed ("institutionalized") into rules of thumb, in proverbs, in local knowledge blocks and in local norms and values that specify how "good farming" is organized. This enormously helps to reduce transaction costs (Saccomandi 1998; Ventura 2001; Milone 2004).
8. In 1966, when the work of Chayanov was published for the first time in the English language, exactly the same questions were being widely asked. Not in relation to agriculture but in relation to the turmoil in southeast Asia, where a peasant army (the Viet Cong) was starting to successfully fight the most powerful army of the world (which it would ultimately defeat).
9. Commons are commonly owned and jointly used resources (or "common pool resources" as Ostrom [1990] calls them) that are used to create value.
10. "The entire future of our country ... depends on the rapid and energetic progress of our agriculture and especially whether or not it is able 'to cultivate two spikes of grain wherever just one spike is growing now'" (Chayanov 1988: 154).
11. "Everybody agrees that the peasant farm is to be the basis for the construction of a new agriculture in Russia" (Chayanov 1988: 137).
12. See also Hardt and Negri (2004), page 123 and note 43.

2

The Two Main Balances Identified by Chayanov

This chapter presents a microlevel analysis of the peasant farm and family. This is not to deny the importance and relevance of the macro level. On the contrary. But there are several good reasons for putting the micro level (i.e., the single peasant farm and peasant family) centre stage. First, because many of the contradictions, relations and trends that characterize the macro level are also expressed at the micro level, often in their crudest form (Mitchell 2002). Second, because the micro level is where the seeds of struggle and change germinate and take root. Third, one of the biggest pitfalls in agrarian studies occurs because of the direct link that is often made between "macro causes" and "macro effects." This frequently applied line of reasoning critically ignores the micro level: the place where trends, predictions, price relations, changes in agrarian policies or any other macro cause are actively interpreted and translated by farmers (and other actors) into a course of action, thus creating the macro effects that actually occur. It's like a process of filtration, with stimuli (prices, policies, etc.) from the macro level always being mediated by and through the actors operating at the micro level. Without understanding the reasoning of these actors it is not possible to understand or predict the effects or outcomes of these macro stimuli. One well-known example is the "inverted supply curve."[1] Chayanov recognized the danger of this methodological pitfall: "to make clear the general economic processes ... we must fully elucidate to ourselves the work mechanism of the economic machine [i.e., the peasant farm][2] which, subject to the pressure of national economic factors, organizes a productive process within itself and, in its turn, with others like it, influences the national economy as a whole" (Chayanov 1966: 120). This methodological stance helped him to avoid deterministic pitfalls.

The Peasant Unit of Production: No Wages, No Capital

Chayanov's analysis starts from a simple but powerful point of departure. Peasant agriculture is (with a few exceptions) reliant on nonwage labour. Labour is not mobilized through the labour market. It is family labour: on-farm labour provided by the farm family. While this seems simple and self-evident, its consequences are far-reaching. Since no wages are being paid, profits cannot be calculated. Consequently, the ordering principles that govern the capitalist economy (e.g., profit maximization and cost reductions that are frequently achieved by reducing the labour input) do not apply to peasant agriculture. Hence, the dynamics of the peasant farm are characterized and governed by a search for internal balances that follow a different rationale.

The difference between the gross product (obtained through commercializing the farm's produce) on the one hand and material expenditure required during the course of the year on the other is referred to as the labour product (or sometimes the family labour product). This is identical to what today's studies refer to as "labour income." It is the income that results from the work done. This labour income or labour product is the only meaningful "category of income for a peasant or artisan labour family unit, for there is no way of decomposing it analytically or objectively" (ibid.: 5). Since no wages are paid, the category of net profit is also absent. "Thus, it is impossible to apply the capitalist profit calculation" (ibid.).

Within the peasant economy, labour is mostly provided by the family. This means the labour market does not govern its allocation and remuneration. The same applies to capital (although this aspect was not explicitly addressed by Chayanov). Every peasant farm contains, and thus represents, capital. But it is not capital in the way that capital is understood in the Marxist sense: as a relation. The "capital" contained within a peasant farm consists of the house and other farm buildings, the land, the many improvements made to it (roads, canals, wells, terraces, increased soil fertility, etc.), the animals, the available genetic material (seeds, a sire), the machinery, the available traction power (of whatever kind). Memory is also an intrinsic part of this capital, just as networks (for selling the products, obtaining mutual help or exchanging seeds) and savings (money available for whatever purchases are needed) are part of it. But this "capital" is not used to produce surplus value to be invested again in order to produce more surplus value. It does "not conform to the classical [Marxian] formula, M - C - M + m" (1966: 10).[3] Nor is it accrued through the exploitation of others' wage labour. In peasant agriculture, "capital" is simply the sum of the available buildings, machines and the like. "By putting a value on buildings, livestock and equipment and by summing these valuations, we can obtain the size and composition of fixed capital for Russian peasant farms" (ibid.: 191). In the family farm, capital is "family capital," which is what most farmers call it. It is part of the resource base created and controlled by the peasant family. It has, first and foremost, a use value: it allows the peasant family to engage in agricultural production and thus earn a living.[4] This "family capital" represents a patrimony. The family tries to extend this patrimony through its life cycle. This can allow them to adopt processes of production that require less drudgery and render more utility (see below). It also functions as a buffer (i.e., an insurance fund) against bad harvests, diseases and the like and, finally, it helps the next generation to start their own farm(s).

The development and use of family capital is not governed by the capital market. There is no intrinsic need to produce a rate of return that equals the average profit rate. Even if the (hypothetical) rate of return is negative, the peasant farm is able to continue its operation and to enlarge its patrimony. The reason is simple: the patrimony does not have to yield any profits. Its value does not reside in the capacity to do so — instead it resides in the fact that it allows the peasant family to make a living, both in the short and the long term. Its use is not governed by the capital market, but by a script defined within and by the peasant family.

It is important to emphasize that the characteristics discussed above (labour being family labour, capital being family capital and income being calculated as labour income) are not limited to traditional agriculture or to remote places. They are also present within current European agriculture. Most farms throughout Europe are family farms, based on family labour and on a patrimony that has often been developed over several generations. This implies, both theoretically and practically, that these units of production cannot be understood as enterprises whose development is directly and exclusively governed by the markets. A good, albeit indirect illustration of this is that in northwestern European agriculture, what is known as the "net farm result" (a virtual concept that calculates the net profit that emerges if all labour had been paid labour market rates and all interest on all capital paid at the current market rate) for most single farms — as well as for the agricultural sector as a whole — is nearly always negative. Not slightly negative, but strongly negative. Hence, these farms cannot and do not function as capitalist enterprises. It would be completely impossible. The explanation is that most "capital" does not have to render the average interest rate. Rather, the available capital represents the resources needed to independently produce an income. The same applies to labour, which is used to satisfy the (many) needs of the family (directly or indirectly) and is also geared toward capital formation ("building a beautiful farm," as I will discuss later). In all these respects the strategic behaviour of the farmers, the way they regulate the different balances entailed in both the farm and the family, is decisive.

One cannot analytically separate the farm from the family to which it belongs and vice versa. Understanding them involves a thorough exploration of the distinct balances that are operated within the family and the family farm. While these balances are operated within the family, their concrete operation extends beyond the family. They link the farming family and the farm unit to the wider environment in which they operate. I will try to illustrate this through an analysis of value flows and, more precisely, how these value flows are socially defined. The first example regards rice production in Guinea Bissau, a country where I worked in the second half of the 1970s.

As exotic as this example might appear at first, we should not forget that the social (as opposed to market) definition of value flows is not limited to faraway and less developed places, such as southern Guinea Bissau. Box 2.2 gives a brief overview of the use of machinery in Europe. The associated value flows are strongly governed by different balances that are informed by farmers' social values. They help to avoid the structure of the farm and the process of production being ordered or governed directly by commodity relations.

2 / THE TWO MAIN BALANCES IDENTIFIED BY CHAYANOV

Box 2.1 The Granary

Rice is the main crop in southern Guinea Bissau. It is grown in tropical rice polders, locally called *bolanhas*. These are beautiful and often very extensive fields, protected by dikes and irrigated with sweet water from the surrounding hills. These fields often produce astonishing yields. The Balanta people have mastered the technique of constructing these *bolanhas* and producing bumper harvests. They use labour groups (a central element in the cultural repertoire of the Balanta) for both construction and production. After the harvest, the rice is collected in huge granaries, locally referred to as *bemba* or *'n ful*. Each extended family (*morança*) has one granary (or a central granary and a set of "satellites"), which is controlled by the head of the extended family. For an outsider the *bemba* only contains rice. But for the actors involved, the contents represent a complex whole of different sources and flows of rice that express different obligations, different destinations, etc. As illustrated in the figure below, the *bemba* is the place where many flows, relations and underlying balances come together and are carefully coordinated in relation to each other.

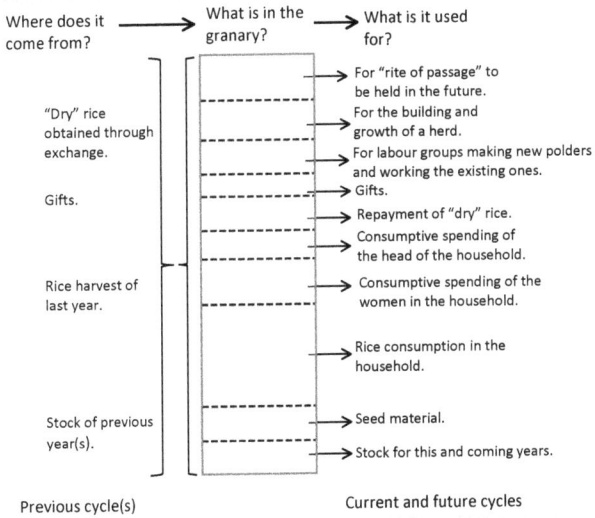

In Balanta society many relations need to be balanced. In the first place these are the relations between past, present and future. Stocks are a strategic expression of this: they are used for food security in both the short and the longer run. In this respect the Balanta are like Chinese peasants. They will only sell (the remnants of) the previous harvest when the new harvest is secured. But the building of extended herds of cattle and savings for the *fanado*, the rite of passage that makes young boys into men, are equally important expressions of the balance between past, present and future.

Second, there are the relations with others. These include the Beafada, a neighbouring people that produce "dry" (non-irrigated) rice, which is much preferred by the Balanta when they have to do the heavy work of land preparation and transplanting. This fresh, "dry" rice gives them extra energy. They receive it from Beafada families and have to "pay back" a corresponding amount of their own rice after their own harvest. Then there are gifts to relatives in the city (who often give gifts in return) as well as gifts exchanged within the village. All this involves maintaining careful balances.

Third, there are the balances entailed within the extended family itself. Part of the rice is for direct consumption, other parts are sold or otherwise exchanged in order to obtain consumption goods (clothes, batteries, radios, bikes, guns or whatever) that cannot be produced in the village itself. Within this category there is a clear distinction between consumptive spending by the head of the extended family or household and the others, notably the women. If the balance is lost here, the women will run away.

In the fourth place, the relations between production and reproduction (notably including the maintenance of the *bolanhas*) need to be carefully regulated. If these get out of balance, an irreversible degradation might occur.

Most of the rice contained in the granaries is sold (this applies to seven of the sections in the figure above). The circulation of the obtained money, though, is strictly limited to specified objectives and destinations. What we witness here is a socially defined process of distribution that is extremely flexible. Changing insights by the actors involved, and negotiations between them, may lead the position of the dotted lines to shift. There is also much mutual interdependence. For example, a reduction in spending on consumption in one year could be used to enlarge the contributions for labour groups, thereby considerably increasing future harvests. In Chayanovian terms this would represent a shift in the balance between drudgery and satisfaction.

Although these value flows are socially defined, this does not imply that the patterns of production and distribution are immune to influences from outside society or from history. On the contrary, at the beginning of the twentieth century, taxes and forced labour induced a strong decline in rice cultivation, which was only countered when a considerable part of the Balanta population escaped from the direct control of the Portugese colonizers. They moved to the empty, nongoverned spaces of the south and made rice cultivation blossom again. Currently, cashew cultivation and imports of cheap rice from southeast Asia are threatening to trigger another collapse in production.

This general pattern guides many other specific practices and relations. Many peasants in the Netherlands and Italy (countries where I have also worked for a considerable time) will, for instance, exclusively link the selling of heifers or tomatoes to the acquisition of feed and fodder for the dairy herd, for example. Thus, when the required additional feed and fodder enters the stable it is "already paid for," as peasants like to say. Through this mechanism the farmer avoids the market becoming the ordering principle for the stable. The social definition helps to keep the market "away" from the stable. Thus, dairy production is de facto distantiated from the market.

In contrast to a capitalist farm, the process of production within the peasant farm is not ordered by the logic of wage labour and capital relations. If profit was the aim people would surely sell their land. But instead, they stick to it, working it or leaving it idle, thus producing a range of unexpected and often counterproductive effects at the macro level (see box 2.3). In short, the labour process, the use and development of patrimony and, especially, the relations between patrimony and labour are not governed by general capital-labour relations. They might be affected by such relations, but they are not directly shaped and reshaped ("determined") by them. The development of the production process might even go against the logics entailed in these general capital-labour relations, just as it might

Box 2.2 Flows of Machinery

There is considerable heterogeneity in northwestern European agriculture. This is often described in terms of farming styles (see chapter 4 of this book). Each style is characterized by strategically ordered relations with, for example, upstream markets, such as the one for agricultural machinery. In specific styles (e.g., "vanguard farmers"), the operators frequently purchase the newest tractors and machines and restructure their farms according to the possibilities offered by these new technologies. Often they sell these tractors and implements after four years (the legally prescribed period for depreciation and fiscal benefits) and acquire the newest ones. In other styles (e.g., "economical farmers," who hate to spend too much money), farmers prefer to buy the second-hand machines being sold by the vanguard farmers. In this way they get them far cheaper and they can use their well developed mechanical skills to maintain them and use them for, say, another twelve years. This allows them to maintain cost levels far below those of the vanguard farmers. Thus machinery flows in particular ways (from industry and dealers to vanguard farmers and then to economical farmers). These flows of machinery (just as the flows of rice in Guinea Bissau) follow specific pathways, through "nested markets" that are defined by the interlocking balances developed by different farming styles. For instance, the balance between capital formation and labour (which is very much a concrete expression of the more general balance of drudgery and utility) differs significantly between the two farming styles.

There are other ways to make machinery flow that correspond with the balances entailed in specific farming styles. These include machine cooperatives, hiring in contractors (often other farmers) with specific machinery or, of course, patterns of mutual help based on reciprocity.

go against the bounded rationalities of the different arenas within which these general relations are embedded (e.g., the markets for labour, capital or food).

All this was very clearly expressed by Chayanov when it came to the controversy about the relative merits of small and large holdings in agriculture. At that time there had already been "a thirty years long polemic ... about the size of agricultural holdings that allows for agricultural development," a polemic in which, as Chayanov explicitly notes, "the works of W. Iljin" (i.e., Lenin) played a major role (Chayanov 1923: 5). According to Chayanov this debate was (and is) grounded on misunderstandings. Size is, as such, not the decisive factor. There is, instead, a historically moving balance of technological development (that allows for larger holdings, although always with a clear upper limit) and the characteristics of the units of production that define a socio-economically optimal size. But these are considerations of secondary importance. "If you want to specify the essential problem, then you should not simply oppose quantitative characteristics of large and small holdings. The challenge is, instead, to analyse, in qualitative terms, the nature of two different economies: the capitalist one and the [peasant one]"[5] (Chayanov 1923: 7). Hence, size is an ambiguous category. What could be large for a peasant farm can be small for a capitalist farm. It can even be too large and too small. It's relative. This also explains "why we do not notice in our surroundings [that covered large parts of Europe of that time] a disappearance of small peasant units. On the contrary, their rank and file increased considerably. The reason for this resides ... in their socio-economic specificities" (ibid.: 6). Further on, Chayanov argues that these specificities, which he synthesized in his theoretical work, compose "the sufficient and satisfactory answer to the question why and how the small peasant farms have historically proved to be able to resist the large-scale capitalist enterprises in agriculture" (ibid.: 8).

The internal mechanics of peasant units and capitalist farms are different. The search for a high rate of return on invested capital explains why the capitalist enterprise is mostly large-scale and seeking to continually expand. Being basically dependent on family labour explains why the peasant unit is mostly small, although historical origins and/or severe marginalization may well also play a role here.

The internal mechanics of the peasant farm, and the associated scripts for resistance and development, are to a considerable degree rooted in two balances (between labour and consumption and between drudgery and utility) that are explored in more detail below.

Box 2.3 Patrimony in the Mediterranean

In Mediterranean Europe the desire to maintain the patrimony (to keep the property in the family) is a basic drive that explains the presence and continuity of many farms (small holdings as well as many of the larger farms), the existence of which cannot possibly be solely explained by reference to markets. These farms belong to pluriactive families: families that gain their income through a multiplicity of activities, farming being just one of them. The family members, to echo Karl Marx, work the fields in the morning, teach in the local school in the afternoon and (maybe) write poetry in the evening, whilst drinking the wine provided by their own vineyard.

Sixty-three percent of male farmers in Italy between the ages of forty to fifty-five only work part-time on their farms. Data from the 2007 census indicated many more have partners who earn an additional income elsewhere. Only 15 percent of these part-time farmers derive all (or nearly all) of their household income from the farm. For 43 percent the contribution that the farm makes to household income is marginal. Moreover, 22 percent of these farms are only viable because part of the income gained elsewhere is transferred into the farm.

These farms are not necessarily small ones. Nor can such constellations be described as the outcome of irrational behaviour. The point is, again, that these farms do not represent capital (in the Marxist sense). There is no imperative need to deliver a predefined rate of return. And the labour used is not wage labour (to be paid according to the standards that reign in the labour market).

Because of the low prices for farm products many of these farms are partly deactivated. This has a negative effect in the regional economy, the landscape and local ecosystems.

The Labour-Consumer Balance

The beating heart of every peasant unit of production is, according to Chayanov, the labour-consumer balance, i.e., the relationship between a family's demands for consumption and the labour force existing within the same family. "For us, the farm family is the primary initial quantity in constructing the farm unit, the customer whose demands it must answer and the work machine by whose strength it is built" (Chayanov 1966: 128). Within this particular balance, labour refers to the available family labour force (i.e., the hands able to do the work) and consumption refers to the mouths that are to be fed. In the narrowest sense, labour refers to the production of food and consumption to eating the food produced. More generally, the balance is about the total production (including that sold on the market) and consumption to meet the many needs of the family, many of which are satisfied through the markets (and paid for with the money earned through production). To be clear, in today's world, just as in the past, it is impossible to reproduce the family and the farm without recourse to the markets. Nobody is independent from commodity circuits. *Robinson Crusoe* was fiction, not reality. All the same, families and farms can relate to commodity circuits in very different ways (see chapter 4).

Labour and consumption are different, incommensurable entities. But they need to be brought into a balance. One implies the other and vice versa. Without consumption there would be no labour. And labour would be pointless if there was no consumption. But there is no simple linear relationship between the two. They are not simply exchangeable. Instead, labour and consumption[6] need to be combined into a dynamic balance that in turn regulates many of the concrete features of the farm and its operation. In early Russia this was particularly evident in the acreage cultivated by each farming family: "the peasant farm in the course of decades ... constantly changes its volume, following the phases of family development, and its elements display a pulsating curve" (Chayanov 1966: 69). The more mouths that need to be fed by a given number of hands, the larger the area cultivated. In situations of land scarcity, the same shift in the consumer/worker ratio translates into intensification or an expansion of "crafts, trades and other *non-agricultural* earnings" (ibid.: 94; italics in the original).

The labour-consumer balance is not the only factor that governs acreage and/or yield levels and is far from being a deterministic factor. Chayanov is quite explicit in this respect: "the family *is not the sole determinant of the size of a particular farm*" (1966: 69, italics in the original). Chayanov probably starts his exposition by discussing the labour-consumer balance for didactical reasons — he subsequently mentions many other additional and/or mediating relations and balances. Together they flow into what Chayanov terms the "organizational plan of the peasant farm." This is an interdependent whole: "Not a single element in the family farm is free; they all interact and determine one another's size" (ibid.: 203). It is interdependent because it is a well-equilibrated whole or, as Chayanov said, in his now somewhat outdated words, a well-equilibrated "economic machine" (ibid.: 220).

The Political Relevance of the Labour-Consumer Balance

To operate successfully the labour-consumer balance on a farm critically needs to meet three conditions.

1) The peasant family needs to receive a proportionate and acceptable share of the overall value it produces. Any increase in their efforts should translate into an improved income. In short, labour needs to provide an income that those involved in the labour process consider to be "just" and sufficient to meet their consumption needs.
2) The relations in which the labour process is embedded are to allow for independence and liberty at the place of work. It is only the peasant family itself that knows the exact conditions existing in the farm and the family. Therefore only the family can assess (whether through internal dialogue and negotiation or through patriarchal imposition) the precise nature of the required equilibrium. Equally, only the farm family can assess how much utility is needed and how much drudgery can be tolerated. Chayanov (1924: 5) was very explicit about this point in *Social Agronomy*, noting that we are dealing with "independent

producers, who run their farms according to their own insights and will. Nobody can dispose of their farms, nobody has the right to submit orders to them." And: "No external authority can run the farm ... Only the direct producer himself who has extensive knowledge of the farm, can run it successfully or, if needed, change it in an adequate way" (ibid.: 6).

3) The labour process needs to be built upon an organic unity of mental and manual labour. Those directly involved in the labour process are the same people who make the main decisions (although there might be complex generational and gender conflicts). To put it differently, the labour-consumer balance precludes any external prescription and control of the labour and production process. This also precludes rigid forms of "horizontal cooperation" (the term Chayanov used to refer to state controlled production cooperatives such as kolkhozes).

The enormous relevance of these requirements, and of the underlying labour-consumer balance, came once again to the fore at the end of the 1970s when a small group of peasants from Anhui, China, started a revolt that finally resulted in a landslide summarized by Netting (1993: viii) as "the dramatic resurgence of the smallholder pattern in China after an era of socialist collectivization." The revolting peasants defined their position with the following slogan: "pay enough to the state, save enough for the collectivity and all that's left is ours" (Wu 1998: 12). This slogan reflects the typical desire of the peasantry to construct and maintain overall balances between the peasantry and the state that are experienced as being fair. Only when such overall balances are well-equilibrated can the farming family satisfy their own needs through their own efforts.[7]

The Scientific Relevance of the Labour-Consumer Balance

The theoretical and methodological relevance of the labour-consumer balance as carrier of the family farm's production machine resides in the fact that it makes clear that the farm, its operation and its development cannot be understood as a simple derivative of external relations and conditions — of whatever kind. This is important when discussing, for instance, agrarian politics or transitional processes. The peasant farm is structured through strategic behaviour that assesses the required balances and then orders the farm and its dynamics so as to meet these equilibria as closely as possible. External relations and trends are interpreted and actively translated into on-farm practices. The peasant farm is, in today's terminology, a smoothly functioning "actor network" that skilfully combines land, plants, cattle, manure, seeds, buildings, labour, crafts, knowledge, machines, networks (and maybe forestry plots, or gardens with medicinal herbs, or agro-tourist facilities or a farm shop). It is an actively constructed response to external conditions, opportunities and threats. This does not apply only to the farm and the way it is operated. It also applies to its dynamics, i.e., the way in which it is actively unfolded.

Understanding the family farm as a well-equilibrated economic machine that is in line with major balances located in the family also negates the view of the peasant farm as an intrinsically unstable system, built upon a contradictory combination of capital and labour.

> Marx had termed the peasant who hires no labour as a kind of twin economic person: "As owner of the means of production he is a capitalist, as worker he is his own wage worker." What is more, Marx added, "the separation between the two is the normal relation in this (i.e., capitalist) society." According to the law of increasing division of labour in society, small-scale peasant agriculture must inevitably give way to large-scale capitalist agriculture. (Thorner 1966: xviii)

Many other Marxists bluntly rejected the conceptualization of the peasant unit as being doomed to extinction. Rosa Luxemburg (1951: 368) wrote,

> it is an empty abstraction to apply simultaneously all the categories of capitalist production to the peasantry, to conceive of the peasant as his own entrepreneur, wage labourer and landlord all in one person. The economic peculiarity of the peasantry ... lies in the very fact that they belong neither to

the class of capitalist entrepreneurs nor to that of the waged proletariat, that they do not represent capitalistic production, but simple commodity production.

A well-balanced actor network can be only constructed if there is a clear strategy that centres on well-specified objectives. What, Chayanov asked, is "the force binding together all the elements of this system" (1966: 103)? This, of course, is the search for an improved family income. It is as simple as that. Yet this very simplicity highlights two major points that have helped to shape the world as it exists today. First, the place of production is the location where the peasant family struggles for its emancipation (materialized through improved incomes, which in turn help to improve the farm). Second, this struggle results in ongoing increases in agricultural production. Consequently, the search for emancipation is the main and decisive driver of agricultural production.

The central role of the struggle to improve income is illustrated by the strong correlations that exist between the income gained from farming and many structural farm features, such as the area sown, the value of the buildings and equipment, the number of cows and animals for traction, etc. (see table 3-18 in Chayanov 1966: 103). "The peasant family, seeking the highest payment per labour unit" (ibid.: 109) develops the farm (i.e., augments the sown area, the number of cows, oxen and horses and invests in capital formation) in order to produce a better income — and the more it succeeds in doing so, the better the family income will be. Elsewhere Chayanov writes "it is obvious that the larger its annual product, the easier it is for the family to find from it the means for capital formation" (ibid.: 11).

However, this cycle is subject to limitations, which are sometimes severe. In the first place it is limited by the available family labour. This implies that labour intensity (the amount of labour invested per unit of land) is bounded. Second, capital intensity (the amount of capital per unit of land) is also bounded: it cannot go beyond the levels implied by available technologies nor can it go beyond the family's possibilities of capital formation. Consequently, the input of both labour and capital depends on another balance — that between utility and drudgery.

The Balance of Utility and Drudgery

This is the second balance discussed by Chayanov. Utility and drudgery are, again, two incommensurable phenomena that need to be brought into a particular equilibrium in order for the peasant farm to function. Drudgery refers to the extra efforts required to increase total production (or total farm income). Drudgery is associated with hardship, long working days, sweating under a burning sun (and dreaming of a cold glass of beer), pre-dawn starts and working under freezing or sodden conditions. Agricultural work might very well be experienced as a joy and a meaningful activity. However, it also involves physical exertion, and when the work to be done increases, its strenuous nature will be more strongly felt. This is what the analytical notion of drudgery seeks to capture. Utility is the opposite of drudgery — the extra benefits (of whatever nature) provided by increases in production. The central point here is that the farming family seeks a balance between the two.

Generally speaking, a growth in production implies an increase in drudgery and a decrease in utility. However, "it would be naïve to consider their link a one-sided dependence of one on the other" (Chayanov 1966: 198). Instead "we have before us two interconnected groups of phenomena which form a single system by establishing an equilibrium between the components of both groups" (ibid.).

The peasant, "stimulated to work by the demands of his family, develops *greater energy* as the pressure of these demands become stronger; ... this brings an increase in well-being" (ibid.: 78; italics in the original). In other words, when the number of consumers per worker increases, the workers' output needs to be higher (for example, by working more land per worker, improving the quality of the resources and/or creating more capital goods). It is here that the balance between drudgery and utility emerges as being strategic. "The energy developed by a worker on a family farm is stimulated by the family consumer demands" (ibid.: 81) and, on the other hand, "energy expenditure is inhibited by the drudgery of the labour itself" (ibid.).

At first sight the labour-consumer balance and that between drudgery and utility seem to be one and the same (especially if one equates drudgery to labour and utility to consumption). While the two are related, they are far from identical; there is a basic difference. The labour-consumer balance relates to the level of the household — it is about the number of consumers in relation to the number of workers. The balance of drudgery and utility refers to the individual worker (and especially to the head of the household): "the greater the quantity of work carried out *by a man* in a definite time period, the greater and greater drudgery *for the man* are the last marginal units of labour expended" (ibid.: italics added).

This difference is strategic, because it explains how the production of the peasant farm can be enlarged and the well-being of the peasant family be improved. By engaging in more drudgery (i.e., by working harder) the single worker(s) can contribute to capital formation, which in turn will allow for higher levels of production with the available labour force (i.e., the net product per worker increases). This subsequently allows for rising family consumer demands to be met.

Figure 2.1 is based on the typical Chayanovian representation of the balance of drudgery and utility. The uninterrupted lines represent "utility" (this diminishes per unit of product as total level of production grows) and "drudgery" (this increases with the further growth of total production). At point E1 the two lines are in equilibrium. This point translates into a level of production (P1). Now, if utility is enlarged beyond the immediate consumption needs of the family (for example, to include the creation of a "beautiful farm" [see box 2.4]), a new "utility curve" is defined, leading to the establishment of a new equilibrium (E2) and, consequently, a new level of production (P2). This then allows the family farm to move beyond satisfying its immediate consumption needs and engage in capital formation (i.e., constructing the ingredients of the "beautiful farm" of the future). Thus, the aspiration for emancipation translates into and occurs through enlarged production and material improvements to the resource base. This might also lead to a redefinition of drudgery; when knowing the act of producing potatoes also opens the possibility of working, in the near future, according to an improved balance, then the drudgery will be felt as less burdensome. Thus a new drudgery line emerges that defines a new equilibrium and corresponding level of production. It is also possible that both utility and drudgery are perceived differently. Then E3 and P3 become possible.

Figure 2.1 Reassessing the Balance of Drudgery and Utility

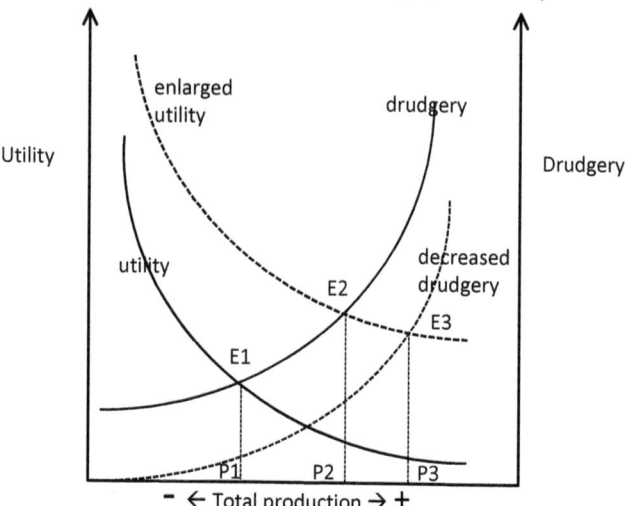

In everyday life, complexities such as the ones shown in figure 2.1 are governed through cultural repertoires (composed of values, norms, shared beliefs and experiences, collective memory, rules of thumb, etc.) that specify recommended responses to specific situations. For example, "a good farmer will never sell his best cow." While this might appear to be a somewhat vague statement, it is, in farmers' daily life, a precise reference to capital formation, to the good offspring that might result from this "best cow." It implicitly states that such a cow is worth the drudgery of caring for it. This is more the case if we know that such a rule of thumb is accompanied by others that specify that, for example, "a good cow is far too risky for a poor farmer" (if it were suddenly to die, the loss would be too great). In short, the active assessments and reassessments of the balances involve judgements based on moral economy (Scott 1976). This moral economy is not external to the "economic machine;" it is essential in making the machine perform (see also Edelman 2005).

There are again several implications related to this particular balance. I will briefly mention two. First, it can be concluded that *the socially and culturally mediated willingness of the peasantry to make agriculture grow and prosper* (to engage in drudgery and to engage in multiple processes of capital formation), *is at the core of agricultural processes of development and growth*. Second, the formation of capital is not necessarily to be organized by, and through, the state (through

Box 2.4 A Current Expression of the Balance of Drudgery and Utility

The figure below (derived from Ploeg 2008) represents a calculus used by farmers from the north of Italy who produce milk that is used to make Parmesan cheese. A calculus is a set of concepts and their mutual relations that is used to specify how farming should to be organized. It represents a particular logic of farming: a particular way of perceiving, calculating, planning and ordering the process of production. The particular calculus outlined here is used by farmers who operate in a peasant-like way. It is not a historic calculus that refers to the past; it is used by farmers currently operating a farm (and who are doing so in a remarkably successful way).

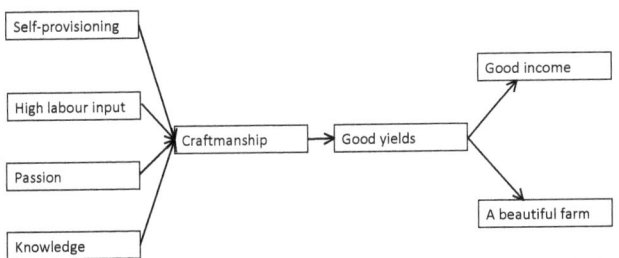

In this peasant logic the notion of *produzione* (good yields) has a central position and significance. Within this logic, *produzione* refers to the production per labour object (i.e., per cow, per unit of land). *Produzione* must be high and sustainable, but, as the peasants argue, it is not to be "forced." It should be as high as is possible within a framework defined by *cura*: care. One has to care well for the animals, the plants, the fields — and if the work is done with care, then production per labour object will be high. *Cura* is also an expression of craftsmanship and refers to the quality of labour. In more general terms, it refers to ordering the processes of production and reproduction in a way that guarantees good yields and steady progress.

In the worldview of Italian peasants (*contadini*), high levels of *produzione* are justified, because they produce and sustain incomes (*guadagno*) in the short run and, probably even more importantly, they allow for the making of a beautiful farm (*la bell'azienda*) in the long run. Taken together, they define "utility" in the Chayanovian sense.

Cura depends on several conditions. There must be *passione* (passion), *impegno* (dedication, which also refers to a high quantity of labour input and to hard work), *professionalità* (knowing the job) and finally there must be *autosufficienza*: the farm unit must be as self-sufficient as possible. The high labour input clearly is an expression of "drudgery," which may be mediated to a degree by *passione* (as illustrated by the shift from "drudgery" to "less drudgery" in figure 2.1).

Overall the calculus shows how drudgery and utility are interrelated in modern dairy farming. The calculus also shows that the balance between drudgery and utility is related to yields. I will return to this point in chapter 5.

an intensified exploitation of the peasantry). It can occur as well as a decentralized process that actively involves the peasant population.

On "Subjective Evaluation"

Over the decades Chayanov's main texts have been subject to much criticism from many different sides. I do not have the space (nor the inclination) to discuss or rebut these different objections here. I will make just one exception. That is the critique that Chayanov's theory on the peasant farm and its dynamics basically depends on "subjective evaluation," that it is not "materialist."

The assessment of the different balances and their translation toward the organizational plan of the farm is indeed subjective in so far as they occur through strategic deliberations and associated "economic calculations" (Chayanov 1966: 86) of the head of the farming family: deliberations that are highly dependent on intragenerational and gender relations. However, the evaluation is also objective insofar as these same deliberations take into account and strongly reflect ("due to necessity," ibid.: 87) the material reality of the farming family (available land, labour force, consumptive needs, need for capital formation, etc.) as well as the structural setting within which it operates (market situation, the possibility of engaging in crafts and trades, price levels, "the influence of urban culture" [ibid.: 84], etc.). The evaluation can even be quantified (ibid.: 87). Subjective evaluation does not imply capriciousness, and/or a disconnection from the material realities of life. On the contrary, it is about taking into account these material realities, which can often be adverse. The point is that these material realities do not impact automatically — they impact through the farmer's active observation, interpretation and translation into a corresponding course of action. All this is done by the grassroots actors who, according to Long and Long (1992: 22–23), are equipped with

> the capacity to process social experience and to devise ways of coping with life, even under the most extreme forms of coercion. Within the limits of information, uncertainty and other constraints (e.g., physical, normative or politico-economic) that exist, [these] social actors are knowledgeable and capable.

Chayanov (1966: 220) himself was aware of the critiques to come: "Due to the use of [such] terms [as subjective evaluation, marginal expenditure and equilibrium],[8] many readers who skim through my theoretical formulas might include me in the Austrian school and thus pay less attention to this study." However, his line of demarcation (and defence) is clear and convincing:

> The marginal utility school [i.e., the Austrian school] attempted to derive from subjective evaluations ... *an entire system of the national economy,* [which] was its main error. I do not do this. My whole analysis ... has been one of *on-farm processes.* (ibid., italics in the original)

He continued:

> I have striven to make clear ... how *from a private economic viewpoint* [today we would say: from the perspective of the actors involved] the family farm's producing machine is organized, how it reacts to the particular effects of general economic factors pressing on it, how its volume is determined, and how capital formation takes place. (ibid.; italics in the original)

Finally, the subjective evaluation is objectively required. Since no wages are paid within the peasant farm; since there is no capital-labour relation to internally structure the unit of production and consumption; and since the required equilibria are not unilaterally imposed from outside, the latter need to be assessed internally, through the subjective evaluation of the involved actors. Such a subjective evaluation is simply indispensable. If there were no such evaluation the outcome would be a chaotic bunch of badly fitting elements (a badly functioning "production machine"). The art of farming is only possible when knowledgeable and capable actors coordinate the many balances entailed within the family and the farm in a tried, tested and goal oriented way. In short, subjective evaluation is intrinsic to farming. It might be the case that the associated marginalist calculation is taboo within particular theoretical strands or among particular political tendencies. But so what? Then we have to adjust theories or redefine the political position. We cannot possibly ask peasants to refrain from making refined calculations and keeping a sharp eye on their interests and prospects. Doing so would be tantamount to inviting them to become the fools of their own village (see also Shanin 1986).

Self-Exploitation

The most unfortunate part of the conceptual scheme developed by Chayanov probably is the notion of "self-exploitation." It has created considerable confusion in the ensuing decades. The term was understood as referring to "excruciating labour by underfed peasant families damaging their physical and mental selves for a return which is below that of ordinary wages" (Shanin 1986). In short, peasant self-exploitation seemed to combine Kautsky's thesis on under consumption (that was presumed to explain the persistence of the peasantry) and Lenin's thesis on the "plunder of labour." Thus, economic backwardness seems to emerge as comprehensive synthesis: peasants are so stupid that they exploit themselves until they are nothing more than skin and bones. They work as hard as the devil, but despite this they can hardly feed themselves.

However, Chayanov himself referred to something completely different, and he was quite explicit about this. "Self-exploitation" equates to the productivity of peasant labour; it is the net product per standardized family worker (Chayanov 1966: 70–71ff.). This "degree of self-exploitation" depends on a range of factors. Chayanov discusses soil fertility, the location of the farm in relation to the market, the current market situation, local land relations, the organizational form of the local market, the character of trading and the penetration of financial capital. A long list indeed that is followed by the remark that all these factors "lie outside the field of our present investigation" (ibid.: 73) — Chayanov confines himself to discussing the factors internal to the peasant household and farm.

The net product per worker depends, of course, on the intensity and length of the work (or the drudgery), other costs involved in production (e.g., seed, tools) and on the remuneration for this work (i.e., the prices paid for the marketed surplus). These prices and costs depend, to a great extent, on the external factors listed above. And here, in my opinion, resides the reason for this strange wording that later caused so much confusion. The concept of exploitation assumes a relationship between two people: one producing a surplus product, the other appropriating this surplus product. Producing a surplus product and then receiving it back simply doesn't make sense. "Self-exploitation" is a self-contradictory concept. One cannot exploit oneself. Again: exploitation assumes a relation, it is impossible at the level of a single and isolated individual. It is even less plausible, since the notion of self-exploitation goes against the core of Chayanov's approach. The capital goods in the peasant farm are not capital in the Marxian sense and no profits (i.e., surplus value) can be calculated. There is just one single return to the family's activity and this return is, by its very nature, unique and indivisible.

In the post 1917 situation, the "current market situation," its "organizational form" and the "character of trading" were strongly influenced by the regime imposed by the Bolshevik state. This regime was heavily exploiting the Russian peasantry, partly to fund the building up of heavy industry. Low price levels, expropriation of parts of the harvest and high taxation all played a role in this scheme. Chayanov (1966b) was very aware that particular economic orders could be superimposed on others. Here the Bolshevik system was superimposing itself on the peasant economy in order to drain it for "primitive accumulation."

Although an important debate about different modalities of accumulation was taking place at that time (Kay 2009), it was probably too dangerous to explicitly discuss "exploitation by the state." Thus "self-exploitation" became the phrase used, suggesting a peasantry that choose to work hard in order to help the building up of state socialism. In reality, though, the term quickly became a slogan for the assumed economic backwardness of peasants (Kautsky 1974: 124 passim). The idea that peasants actually want to be peasants was inconceivable for Kautsky — as was the notion that "it was and is through this 'self-exploitation' that the peasantry produced progress" (Vlaslos 1986: 158).

Notes

1. A "normal" supply curve predicts that price increases will augment production and price decreases will reduce production. But it often happens that African farmers produce less when prices increase and European farmers produce more when prices decrease.
2. Chayanov was fond of using the metaphor of the machine ("the well-working machine," "the economic machine") when referring to the peasant unit of production.
3. Here "M" refers to money, "C" to a commodity acquired with this money and "M + m" to the initial amount of money ("M") increased with an additional amount (or surplus value) equal to "m." Hence money is converted into a commodity and then this commodity (notably wage labour) is converted into more money.
4. This does not exclude, of course, the potential of capital relations "penetrating" into the peasant farm. I will discuss several mechanisms through which this occurs, their impact and their theoretical implications, in

chapters 4 and 5.
5. Chayanov literally writes here (in the German translation) "*und der lohnarbeiterlosen,*" i.e., the one without wage labour. This equals the peasant farm.
6. Here we probably have a flaw in Chayanov's exposition: he does not discuss the possibility that the peasant family actively regulates the consumer/worker ratio by itself (by marrying later, for example, or, as happens today, through birth control). See Hofstee (1985) and Netting (1993: 315), who show the shift over time in the demographic balances of rural societies.
7. Similar relations exist elsewhere. Many rural movements in Europe (e.g., the recent milk strikes) have been driven by a generalized feeling that "the balance has been lost."
8. "These and other concepts ... are so unusual that ... I run the great risk of not finding a common language with the Russian reader" (1966: 219).

3

A Wider Array of Interacting Balances

On the one hand, the wider array of balances I discuss in this chapter relate to the two balances extensively discussed by Chayanov and briefly synthesized in the preceding chapter. On the other hand, this wider set of balances — largely developed within the tradition known as the Chayanovian approach — allows us to come to grips, in a coherent way, with the problems and potentials facing peasant farming today. The same set of balances also helps to explain the considerable heterogeneity that exists among the peasantry, both between and within countries and regions. I will present the balances in what seems to me to be the most logical sequence.

The Balance between People and Living Nature

In its most general sense, farming should be understood as coproduction, that is, the encounter between the social and the natural (Toledo 1990). In this sense, farming can be seen as the ongoing interaction, and mutual transformation, of people and living nature. Humankind uses nature and in doing so transforms nature. But using nature (in particular ways) also puts an imprint on society itself. The transformation of nature requires specific institutions. Therefore, coproduction shapes and reshapes the social as much as it does the natural. This is beautifully expressed in the answer a French winegrower and cooperative leader once gave me when I asked why he referred to himself as "peasant": "*Moi je suis paysan parce que je vive de la terre*" ("I am peasant because I live off the land"). Slightly rephrased, this could read "coproduction makes me a peasant."[1]

"People" and "living nature" are different entities. Yet they are combined in the practice of farming, which involves constructing a proper equilibrium that needs to meet several objectives. It has to provide sufficient production (allowing for "living off the land"). But it also needs to reproduce nature, preferably enriching, improving and diversifying it. Using and transforming nature also implies that people are able to cope with diversity, uncertainty and capriciousness. Those engaged in coproduction have to face unfolding cycles (the development of a crop, the growth of calves into heifers and then into milking cows) and translate their observations back into these cycles, adapting them in many ways, some small, some large. The labour process is, therefore, organized in an artisanal way with manual and mental labour being closely interwoven. In this respect, the existence of external centres of command can only have detrimental consequences (Sennett 2008). Farming needs to be tuned to the specificities of time and place. In *Social Agronomy*, Chayanov (1924: 12) wrote, "working with blueprints is impossible." All this decisively favours the peasant farm as the organizational model: it is the most appropriate institution for managing coproduction. Coproduction excludes standardization, complete quantification and tight planning. Hence, it requires the peasant farm, since the latter ties the well-balanced development of coproduction to the peasantry's emancipatory aspirations. This is done at the micro level of the peasant farm by firmly establishing a direct link between the unfolding of coproduction and the improvement of the family's labour income.

The centrality of coproduction implies a series of far-reaching consequences. It implies in the first place that agricultural development cannot be understood as the more or less perfect unfolding of the fixed laws that are presumed to govern nature and the economy. It is, instead, the outcome of ongoing interactions and transformations that repeatedly create new constellations, each with their own regularities and potentials (see chapter 5). Coproduction implies that nature can be enriched and that new potentials might emerge. Landscapes are formed and reshaped through particular forms of coproduction (Gerritsen 2002); animals, plants, marshes, woodlands, hills and streams are transformed. When the different remoulded elements are recombined this can create new productive possibilities.

Second, the malleability (or more generally, the transformability) of natural resources[2] — such as fields, cattle and "the nature of the countryside"[3] — allows for agriculture to develop endogenously. Growth and development can be produced "from within," as I will argue in more detail in chapter 5.

Third, coproduction (and the possibility of endogenous development) puts skills centre stage. Skills are about being able to "see the bigger picture" — to observe, handle, adjust and coordinate a wide range of domains within the social and the natural worlds and, particularly, their interactions.

Fourth, it is important to recognize that, in peasant agriculture, the balance between people and living nature is essentially one of reciprocity (see box 3.1).

Through coproduction and co-evolution both the social and the natural are continually transformed. Chayanov was deeply aware of

Box 3.1 On the Reciprocity of Man and Living Nature

When Italian peasants discuss the way they relate to their fields, their cows and their crops they probably will use the word *cura* (see also box 2.4). This expression has strong associations with craft and craftsmanship, but it also refers to "care," just as the verb (*curare*) refers to giving care. It is, essentially, about reciprocity (see Sabourin 2006). It is only when taking care is central to labour that the land, the animals and/or the crops will render you a good yield. Giving care is far from just an instrumental activity. It supposes, within the discourse of Italian peasants, the presence of passion, commitment and knowledge about your objects of labour. Finally, there is the requirement of self-provisioning: the resources used in the process of production should be owned by the farming family itself. Tight dependent relations with markets at the input side of the farm are to be avoided, because they would bring "the logic of the market" into the heart of the farm. This would threaten, if not exclude, working with *cura*. The concept of *cura* defines, and simultaneously reflects, a reciprocal relationship between the farmer and his objects of labour. This relationship is definitely not a commodity one. It is about giving and getting back. It is, as it were, about gifts that flow in two ways. The farmer raises and takes care of the calf, gives her shelter and the opportunity to develop into a good milking cow, and then he will feed her, probably with a diet that is carefully adapted to her individual needs. In return the cow will give the farmer new, hopefully promising, calves and a rich flow of milk that might continue for many years. It is, as Victor Toledo

(1990) would put it, a noncommodity exchange between farmer and living nature.

This type of relationship underlies many farming systems around the world. According to van Kessel (1990: 78) an anthropologist who worked for many decades in the Andean region, this reciprocity is strengthened through "metaphoric connotations" that imply a kind of personification: land, crops, lakes, wells, but also the light, the rain, the frost and other meteorological phenomena are perceived and understood as living beings that give all kinds of signs. In this context it becomes nearly self-evident to say, for example, that "this piece of land is grateful" (for all the care it received) and that consequently "she (land is almost exclusively feminine) is generous" (i.e., willing to give back). Equally telling is the use of the subjunctive. When talking to (or about) the objects of labour, Andean farmers do not refer to the world as it is (a world that is given once and for all and governed by mechanistic cause-effect relations) as would be the case when the indicative mood would be used. Instead, the subjunctive refers to possibilities, to evolving realities and to expectations. It reflects intuitions. This does not imply that Andean peasants are dreamers — on the contrary: "the norms for technical operations in the field are dedication [*compromise*], understanding [*comprensión*] and affection [*cariño*]" (van Kessel 1990: 92). These concepts coincide strongly with the Italian notions discussed above, just as the Frisian saying "*as jo lân hâlde wolle dan moat it sines ha*" ("if you want to stay on the land, you have to give it what it needs") echoes the same give and take relationship (Ploeg 2003: 94). Such similarities are far from coincidental. They are rooted in the reciprocal relationship between people and the land and thus emerge wherever peasant farming is practised. As the Chinese proverb goes, "If man works hard the land will not be lazy" (Arkush 1984).

this. In the second part of his *Economy of Labour* he notes, "The peasant economy of 1917 is not anymore the one of 1905. The peasant economy itself has changed deeply: the fields are worked differently and cattle are raised in a new way. The peasants sell more and they also buy far more. Cooperation has been extended considerably in our countryside and thus deeply changed its *nature*. The peasants themselves [have] progressed a lot and became more civilized" (Chayanov 1988: 136; italics added).

Chayanov did not elaborate this balance. This is quite understandable. As discussed before, in the late nineteenth and early twentieth centuries there was hardly any scarcity of land in Russia, and land transfers were common due to the repartition of commune lands and to widespread renting. This meant that further growth could be achieved by simply rolling the existing patterns of land use over larger areas. There was less need to intensify farming (and if intensification occurred it was mainly through changes in cropping schemes). In labour driven processes of intensification, resources are continuously improved, through fine-tuning the way they are used and combined, in a constant search for ongoing progress. Hence, the interaction of people and living nature is continuously being reconstructed. The moves from one level of intensity to the next highlight that farm practices are socially constructed and underline the continual transformation of both social arrangements and ecological patterns — sometimes in ways that are slow and barely visible, at other times abruptly. It is telling, in this respect, that Chayanovians such as Vries (1931) and Timmer (1949), who both worked in Indonesia as well as in the Netherlands, paid considerable attention to this particular balance. What was hardly visible in Russia was extremely prominent in the locations where they operated.

The balance between people and living nature is the first that needs to be considered in any analysis of contemporary agriculture. This is due to the many disconnections that have been created between farming and ecology, which have resulted in an accelerating environmental crisis.

Achieving the right balance between the social and the natural is an ongoing concern in all farming practices. Sometimes farming moves away from living nature, at other moments it re-grounds itself upon it. Jozef Visser (2010) documented an important episode in the immediate aftermath of World War II, when war machinery was transformed to be used for other purposes. Thus, ammunition factories were converted into factories to produce chemical fertilizers (relatively easy because both are based on the Haber-Bosch process). Production lines for armoured vehicles were geared to making tractors. Much of the repressive legislation that had subordinated farming to the needs of warfare remained in place, on both sides of the line separating the combatants. Marshall Assistance was used to provide agrarian sciences with a new agenda that reflected the "entrepreneurial farming" that had developed in the U.S., which differed greatly from the peasant farming that dominated Europe. Soil biology and the focus on maintaining soils rich in biological life that could deliver nitrogen naturally disappeared from the agenda, to be replaced by soil chemistry. And finally, the "science" of logistics, which had developed enormously during the war, was applied from the mid 1950s onward to plan and bring about the so-called modernization of European agriculture — a campaign that was repeated in large parts of Asia and Latin America through the Green Revolution.

Modernization and the Green Revolution represented an important rupture from farming as the coproduction of people and living nature. Chemical fertilizers took the place of soil biology, manure and peasants' knowledge. Industrial concentrates replaced meadows, pasturelands, grass and hay. Natural mating disappeared, whilst artificial insemination and, later, embryo transfers and computerized selection of the best sire, started to dominate. Electrical lighting has replaced sunlight in much of today's horticulture, whilst in chicken sheds a twenty-four hour period now entails two nights and two days, in order to accelerate their growth. Solar energy became less important and was increasingly displaced by fossil energy. All this is indicative of a decrease in the role of nature, more so if one takes into account the re-engineering of what remains of nature through, for example, genetic modification. But further steps are still possible. For instance, large-scale entrepreneurial dairy farming in the U.S. is currently "rebuilding" living nature in a remarkable way. Veterinarians working for these large "milk factories" systematically remove the cows' uteruses after the first calving. This is done in order to standardize the hormonal cycles, which would otherwise fluctuate strongly when the cows come into heat, deliver calves and begin and end their lactation cycles. Such fluctuations require frequent adjustments to the feeding regime, which is at odds with the standardized management of large herds in capitalist farm enterprises. So instead, the uteruses are removed, the animals receive frequent injections of the BST hormone in order to continue milk production, and they often collapse after some one thousand days of production. The *animale tecnologico* ("technological animal") as my Italian former colleague Ballarini (1983) called it, is becoming a new reality that is at odds with both nature and the ethics of society (and therefore is kept well hidden). Cloning, in vitro fertilization and food engineering are other examples of the subjugation of nature to the requirements and interests of large-scale agriculture and food production enterprises.

There are many counter movements as well.[4] We can refer to organic agriculture, low external input agriculture (Adey 2007), the style of farming economically (Ploeg 2000; Kinsella et al. 2002; Domínguez García 2007; Paredes 2010) and numerous agroecological movements (Rosset and Martínez-Torres 2012). They all propose a far-reaching reshuffling back toward coproduction. In all these approaches living nature once again plays a central and co-ordering role. Thus these counter movements are helping to make agriculture more peasant-like. Simultaneously, they are helping to redirect much of agronomy toward the "social agronomy" proposed by Chayanov.

The Balance of Production and Reproduction

Farming is not an extractive process (although adverse circumstances might push farming in that direction). Farming entails both production and reproduction. It is grounded on the ongoing reproduction of the resources it uses. This reproduction not only involves "living nature," as discussed in the previous section, it concerns all the resources, all the elements required to make farming function smoothly. Chayanov often referred to reproduction as "capital renewal." Thus Chayanov (1966: 120) points out, "it is clear that the family of the peasant-run farm ... tends in the final result to satisfy its demands to the fullest extent possible and to ensure the further stability of the farm by a process of capital renewal with the least expenditure of energy."

The historical development of the balance of production and

reproduction is extensively discussed by Anne Lacroix (1981). At first, the surrounding ecosystem was used to renew resources. Slash-and-burn agriculture is a typical example: when a field is exhausted, it is abandoned and a new field is taken from nature.[5] Objects of labour and instruments (see box 5.1) are derived from the surrounding ecosystem, whilst the available labour force carries the knowledge about how to use the surrounding ecosystem.

In a second historical period, reproduction shifts to the farm itself. It becomes an integral part of farming: fields are actively fertilized, plant varieties selected, cattle improved and the newly constructed fields, animals and crops become the proud symbols of the relative autonomy that allows peasants to go beyond the often strict limits of the local ecosystems.

In a third period, the current one, reproduction has once again moved away from the farm. It is externalized to agro-industries that increasingly produce and deliver the objects of labour, instruments and the manuals to be followed by the labour force (Benvenuti 1982; Benvenuti et al. 1988). In this new constellation it is no longer the peasant community that builds its "code" into the objects and instruments (as was the case in the second period) — it is now agro-industry that builds a specific and often scientifically designed code into the different artifacts needed on the farm. There can be considerable differences between the two codes. The code that Friesian farmers built into their dairy cows includes the centrality of roughage produced on farm (grass, hay, silage) to feeding the cows. The code of Holstein cattle, one of the main "artifacts" of the powerful breeding institutions that control the trade in semen and, more recently, in embryos, typically reflects the opposite, i.e., the centrality of industrial concentrates. In this way dependency becomes an inbuilt feature.

The balance of production and reproduction is easily broken. An imbalance might be induced by external factors, but the dangers might equally come from inside. The latter is most likely to happen when peasants are looking for short-term advantages. This occurred in Frisian dairy farming in the first half of the nineteenth century. At that time, butter prices were so high that the peasants used all their pasture for milking cows in lactation, to get as much milk as possible to make into butter. Calves and heifers were restricted to the periphery of the farm, to wet pieces of land with little, low-quality, biomass. They received little care. In short, reproduction was neglected and production dominated. Within two decades the result became evident: the quality of the breed was ruined. The animals were far smaller and milk yields considerably lower. This painful lesson became a crucial ingredient of collective memory: "a good farmer is not a merchant" (meaning that building and reproducing a high quality resource base always comes first).

More usually, though, it is a combination of external pressures and internal drives that produces an imbalance. A current example is the degradation of the beautiful tropical rice polders (see box 2.1), first in the Basse Casamance in Senegal and more recently in Guinea Bissau. Low prices for rice (especially due to cheap imports, highly imbalanced government supports and unaccounted environmental costs) have considerably reduced the potential income from rice farming. This, combined with the temptations of the cities (and international migration), implies that most youngsters leave the villages. Hence, the maintenance of the polders (usually done in the dry period) has come to an almost complete stop. This has led to declines in yields and production and, in the end, a possible complete abandonment of the once highly productive *bolanhas*. There are also cases in which external factors have dominated, particularly in Latin America. One example is the credit policies of the *bancos agrarios*, which would give credit for productive activities (although hardly enough, for the most part), but refrained from any assistance to reproductive activities (e.g., maintaining fences) on the grounds that such activities are "unproductive." While true, this view is extremely short-sighted and shows little understanding of the importance of maintaining a balance between production and reproduction.

The Balance of Internal and External Resources

Alongside the resources that are produced and reproduced in the farm itself (the internal resources), every farm, wherever located, also needs external resources. It would be impossible to imagine farms functioning without them. However, the nature of these resources, their origin and, especially, the way they are acquired and the effects of the method of acquisition, can have far-reaching consequences.

For many resources there is a considerable exchangeability between internal and external ones. Cows might be reproduced on the farm itself (selected calves are raised into heifers that, after their first calving, might replace an older milking cow); equally they might be bought at the cattle market. The farm may have its own sire (probably shared with neighbours), but the required semen might equally be bought from a station for artificial insemination.

What applies to cattle also applies to cattle feeding. Hay, grass, silage and protein-rich crops to be added to the rations might be produced on the farm; however, roughage and concentrates might also be acquired on the market. Fertilizers might be bought or produced on the farm itself (examples of fertilizers produced on-farm include "well-bred" manure and nitrogen fixation through clover or alfalfa). Labour might be mobilized through the market but also be provided by the family and/or the local community. "Capital" might be produced on the farm itself (through capital formation but also in the form of savings); it might also be acquired in the capital market. The same applies, to a degree, to machinery. It might be mobilized through different mechanisms that mediate the impact of the market in contrasting ways (see box 2.2). "To make or to buy" became, in the course of the twentieth century, the central question out of which neoinstitutional economics was born. It could be argued that peasant agriculture is a nearly perfect textbook illustration of neoinstitutional economics (Saccomandi 1998; Ventura 2001; Milone 2004)[6] for the balance of internal and external resources is all about the choices between "making" or "buying."

Figure 3.1 synthesizes the technical exchangeability of internal and external resources. It also illustrates the associated flows (Georgescu-Roegen 1982; Dannequin and Diemer 2000). First, figure 3.1 shows that that agriculture is a process of conversion: resources are converted into useful products. The process of conversion is grounded on a twofold mobilization of resources. Some of the resources are produced and reproduced within the farm, others are acquired through the markets. In turn, the process of production generates three flows: a marketable surplus that is sold on the markets,

Figure 3.1 The Flows Entailed in Farming

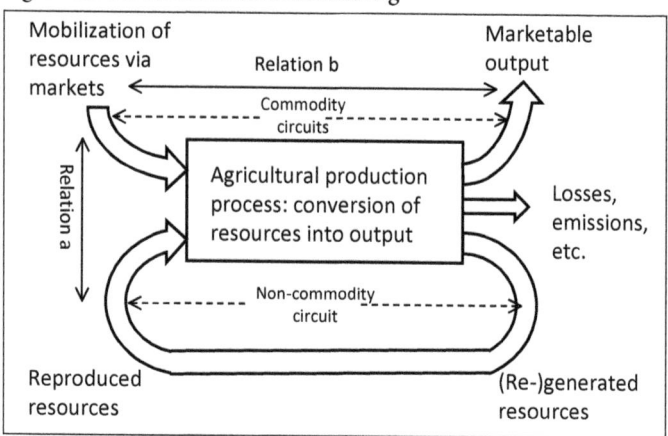

a part that is reused on the farm and the inevitable, although highly variable, losses and emissions. To this we can add that the combined and simultaneous generation of products to be sold on the markets and the products to be reused is partly due to the materiality of living nature: When producing potatoes, there will be potato seedlings as well. When producing milk, there will be a calf (unless the uterus is removed). But of course, the seedlings can be eaten at times of need (or seedlings from "improved varieties" might be bought to replace the available ones). And the calf can be sold later on in order to buy a heifer from a new breed. The important thing is that there is room for manoeuvre that allows for different choices. If the upper left flow dominates over the lower left one (the self-provisioning of required resources) then commodity relations penetrate into the core of the farming process. This leads the farm to become market dependent (especially on the upstream side) and the farm becomes structured as an entrepreneurial enterprise. If, however, the lower flow is the dominant method for acquiring resources, then there is relative autonomy and farming tends to be structured as peasant agriculture. In peasant agriculture the market is, above all, an outlet (on the downstream side), whilst entrepreneurial and corporate agriculture are essentially ordered by the markets and need to follow their logic.

The use of external resources brings opportunities but also

often has highly distorting consequences. This implies the need to repeatedly define and construct a specific and well-thought-out equilibrium of internal and external resources. Relying on external resources can help to considerably reduce the drudgery faced by the farming family. But a farm that is highly dependent on upstream markets might potentially be devoured by these markets. Assessing the right balance[7] also helps to create relative autonomy: a position that allows for styles of farming that suit the interests and prospects of the farming families (see the following section).

This relative autonomy (or, vice versa, market dependency) might be measured as the "degree of commoditization." There are different ways to approach this question. Figure 3.1 illustrates two possible operationalizations: "ratio a" and "ratio b." The latter one (that is identical to labour income as defined by Chayanov) was used throughout the agrarian history of the Netherlands by peasants themselves as the "clean part" (Ploeg 2003). It seems to be a nearly universal yardstick: Chinese peasants of the twenty-first century also calculate using an identical, albeit differently phrased concept (Yong and Ploeg 2009).

Chayanov (1966: 120–1) attributed strategic importance to the degree of commoditization: "Among the [many] differences in the farm's organizational plan, the most basic one which determines the whole character of the farm's structure is the degree to which the farm is linked with the market — the development of commodity production in it." He limited himself to analyzing the output side of peasant farms; some sold most of their produce to the markets, in others most produce was meant for self-consumption and only a small part was commercialized (Chayanov 1966: 121–2ff.). Chayanov (ibid.: 258–263) did not address variations in market dependency on the input (i.e., upstream) side of single farms, although he was well aware of the gradual subservience of agriculture to the wider industrial, trading and banking circuits that started to take shape at the beginning of the twentieth century.

As pointed out in the previous chapter, labour income is, within the Chayanovian analysis, the central source of income within peasant agriculture. As a matter of fact it is, as we saw, "the only possible category of income." Here it is important to point out that this labour product is defined by two sets of transactions. These are, first, all the transactions at the output side of the farm. Together these define the gross product. Note that this gross product is not identical to total production, since part of the production might be used in the farm unit itself. This latter part is illustrated in figure 3.1 as the flow of regenerated/reproduced resources. The second set of transactions is located at the input side of the farm, and it embraces all the monetary expenses incurred (all "material expenditure," as Chayanov phrased it). Thus labour product equates to gross product minus all monetary expenses (in figure 3.1, marketable output minus the resources mobilized on the markets).

A farmer needs to balance these two sets of transactions in a way that will lead to an acceptable labour product. One possibility is to reduce the expenses related to the supply of external resources as much as possible. This can be done by developing and using internal resources to replace the external ones, the approach favoured by the agroecological movements. It is a move that counters the long-term trend that Chayanov already noted in his time. During the last sixty years in particular there has been a marked increase in farms' dependency on external resources. Ironically, this same trend (with its associated consequences) has revitalized the old peasant wisdom that the more independent you are from the markets on the upstream side of farming, the better position you are in with relation to the markets on the downstream side. Thus, alongside the ongoing processes of commoditization we increasingly also witness processes of (relative) decommoditization.

The Balance of Autonomy and Dependence

When assessing the impact of the balance of autonomy and dependency, one has to take into account "the social institutions that surround the production and distribution of wealth" (Little 1989: 118). Whilst the farming economy is undoubtedly "an organized system of social relations and independent decision making" (ibid.: 117), it is simultaneously, through the dependency relations in which it is embedded, subject to surplus extraction. It is here where "class relations and the particulars of an existing system of surplus extraction" enter the analysis, as Little (1989: 118 passim) convincingly argues. To illustrate his point, Little refers to Victor Lippit (1987), who applied a surplus extraction framework to analyze the traditional Chinese rural economy. Lippit demonstrates that this traditional agrarian economy had a sizeable surplus and that the rural elite effectively extracted this surplus from peasants and artisans. "The mechanisms of extraction differed — rent, interests, taxation and corrupt tax practices — but the effect was the same: to transfer from the immediate producers to a small elite class some 25 to 30 percent of total rural product" (Lippet 1987: 120). This created a persistent stagnation; peasants lacked the means to invest and thus develop farming further, whilst the rural elite frittered away the extracted surplus in luxury consumption. Thus, as Little (1989: 118) concludes, "the surplus-extraction model ... directs us to consider the system through which various elements of the elite class are enabled to seize part of the surplus created through productive economic activity. How is this surplus created and by whom?" Associated with this, the direction of agricultural development "heavily depends on the incentives, opportunities and powers conferred on the class parties by the class system; class relations thus impose a logic of development on the system" (ibid.). Here we clearly see that a Chayanovian approach does not exclude class analysis (as is sometimes assumed). Politico-economic analysis (including class analysis) enters as soon we analyze the operation of peasant units of production within the context in which they are located. The same occurs when the analysis starts at the macro level, for example, when asking how a particular politico-economic formation affects rural development. Then a Chayanovian understanding of the peasant unit needs to be included in the analysis because the effects of the particular politico-economic formation are mediated by direct producers who try to assess the important balances within their units of production according to the reigning parameters.

If both sides of the equation are taken into account it is possible to define the peasant condition as a struggle for autonomy and improved income within a context that imposes dependency and deprivation. The context can be analyzed with the surplus extraction model; the actions taken to respond to this context are best understood through a Chayanovian approach. In the concrete analysis the one assumes the other and vice versa.

This is beautifully illustrated in the seminal work of the agrarian historian Slicher van Bath who puts the notion of "farmers' freedom" centre stage. This notion contains two components: the "freedom from" and the "freedom to." The first can be identified by politico-economic analysis, the second further specified by a Chayanovian type of analysis. "Weighed down by the burden of certain expenses and obligations [the peasants of the past] were limited in their actions" (Slicher van Bath 1978: 72). They were not free *from* the many dependency relations and the associated levies, expenses, taxes, etc. Hence, the "clean part" (see above) was limited and this reduced the freedom *to* develop the farm according to one's own interests and prospects: the less the freedom from, the more restricted the freedom to. Slicher van Bath (ibid.: 80) observes that this double freedom "is determined by various factors, which are in turn the effect of historical circumstances." He equally shows that "liberties are nowhere stationary, they are subjected everywhere to historical evolution and digression" (ibid.).

In the same vein, Ernst Langthaler's extensive study of Austrian farming, covering the 1930–1990 period, led him to the conclusion that "the more subordination to factor and product markets gains hegemony, the more class-differentiation between accumulation and proletarization takes effect; vice versa, the more the farm's self-controlled resource base is strengthened, the more the family members are able to cope with unfavourable conditions of the political-economic system in their life-worlds" (Langthaler 2012: 400). He adds: "the resilient family farming system in bureaucratic and capitalist environments resembles a *Stehaufmännchen* [a puppet that always bounces upright when pushed over]; metaphorically speaking, *family farms wobble, but they don't fall down*" (ibid.; italics in the original). They keep (re-)setting the different balances in order to establish, time and again, the needed equilibrium.

The Balance of Scale and Intensity
(and the Emergence of Farming Styles)

In the concrete organization of the farm there is yet another balance that needs to be carefully assessed. That is the one of scale and intensity. Scale refers to the number of labour objects (units of land, animals, etc.) per unit of labour force. Intensity refers to the production per object of labour (for an extended discussion see box 5.1). In an international comparison Hayami and Ruttan (1985) argue that there are two contrasting ways to increase incomes in agriculture. These are intensification and scale enlargement (although, of course, all kind of combinations and intermediate positions are possible).

It is important here to return for a moment to the notion of coproduction. This implies, among other things, that agriculture is malleable. It can be organized in different, contrasting ways. This is important as it allows for the "organizational plan of the farm" (Chayanov 1966: 118–94) to be set according to the needs, interests and prospects of the farming family. This "setting" occurs through tuning the different balances.

Intensity and scale define a two dimensional space (see figure 3.2) within which different positions, i.e., different styles of farming, can be discerned. Within areas with similar ecological, economic and institutional conditions it is nearly always possible to find a range of different styles (or differently tuned machines, to echo Chayanov's phraseology). A few of these are outlined below.

The style of farming economically is characterized by a relatively low scale and relatively low intensity. According to the Hayami/Ruttan model this implies poverty. However, this is not necessarily the case. Rather, this style, which is based on cost reductions, highlights a theoretical omission in the model of Hayami and Ruttan: it does not include costs. The balances in this style are set in such a way that spending on external resources is minimized, while coproduction is prioritized. This reduces dependency and augments autonomy. At the same time, financial costs (related to growth) are minimized. Thus the overall costs are low and labour income is high (also when expressed in relative terms as, for example, labour income per 100 kg of milk). Under crisis conditions this style turns out to be highly resilient.

Figure 3.2 Farming Styles

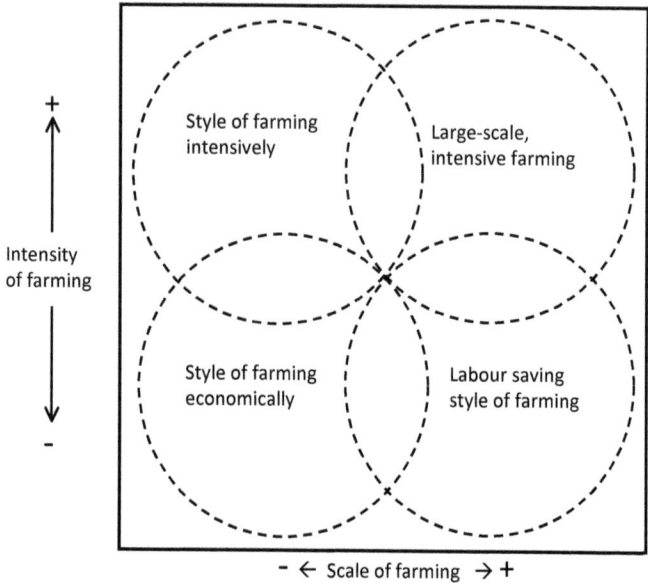

The central aim of farming intensively is high yields (the "good cow" is a typical symbol here). In the labour saving style of farming (for which the "powerful tractor" is the most telling symbol) the aim is to have as many objects of labour as possible and to minimize labour input. Together, these two styles typically make up the hotly debated "inverse relationship" between farm size and productivity. Once, this was indeed the dominating interrelation. Such a relationship is still discernible today, but it is no longer the only one. Apart from the style of farming economically, another style has emerged, that of large-scale, intensive farming. This style is a co-construction of agrarian policies and technological development on the one hand and the strategy of agricultural entrepreneurs on the other. Technology comes in as new scientifically elaborated artifacts such as cubicle stables, Holstein cattle, nitrogen-sensitive grassland varieties and concentrates, which together allow for technology driven intensification that also has the effect of enlarging the scale of farming (see chapter 5). Agricultural policy has an influence by stimulating the creation of big farms (e.g., through investment subsidies, spatial re-organization) and offering long-time security by setting stable prices, as was the case with the earlier Common Agricultural Policy of the European Union. The agricultural entrepreneurs' role in this process is to try to develop farming by taking over resources from other farmers.

In *Social Agronomy*, Chayanov (1924: 2) spells out some of the processes that underlie the production of heterogeneity within farming.

> The individuality of the direct producer, his creative energy, the particularities of his farm and the quality of his fields, mean that the individual farm will always deviate from the average type. Curiosity and the search for novel solutions characterize all farmers. Consequently, all farms are in a kinetic condition; they are permanently changing due to the widely spread experiments, searches and creative trials.

The actively created heterogeneity (condensed here as different farming styles) constantly interacts with the many changes in the context in which farming is embedded. The impact of these changes will have a different effect on farms practising different farming styles. Hence selection will occur; some styles will show themselves to be better adjusted to facing and dealing with the changed environment, others will become marginalized. This creates variation and selection "that is in large traits the mechanism of rural development ... [T]here is no collective will, no overarching consciousness, no commander and no plan" (ibid.: 3–4). The centrality of variation and selection does not exclude the importance of supporting or strengthening the search for the most appropriate styles, which Chayanov thought could be achieved through a "societal ratio" (ibid.: 3). This was precisely the purpose of the social agronomy Chayanov proposed. Such a search that includes the creative construction of all manner of in-between positions and mixtures is, and remains, important.

As Langthaler (2012: 402) concludes in his impressive study of the *longue durée*, "it is the hybridity of family farming styles that increases the resilience of family farming systems in the challenging environment of post-war organized capitalism."

Fighting for Progress in an Adverse Environment

In today's world, the labour-consumer balance takes a very different form from the one described by Chayanov. For the Russian peasants of the first two decades of the twentieth century, the consumption side of the equation mainly (although far from exclusively) came down to the consumption of food, clothes and the like (see, for example, Chayanov 1966: 122, table 4-2), whilst the self-provisioning of the farm went without saying. These were self-evident features, the more so since goods or services that were lacking could often be obtained through socially regulated exchange. The farm produced for the market but could do so because its most immediate needs were satisfied through self-consumption and self-provisioning.

Nowadays, consumption embraces many elements that cannot be provided from within the farm: education, electricity, mobility (at least beyond a certain radius), communications, luxury items, etc. The demands "of the customer that must be answered" (ibid.: 128) have changed significantly. Equally, the running of today's farms requires a range of items (tractors, energy, pumps, etc.) that cannot possibly be produced within the farm itself. "The work machine" has been significantly altered. Together, these changes mean that the labour-consumer balance now needs to take a far wider range of markets into account. The direct relations between labour and consumption are being reduced whilst the indirect relations (which critically assume a combination of several market transactions) are now more important. Assessing the labour-consumer balance now involves deliberations about many markets, their interrelations and expectations about the main tendencies within these markets. The needs of the family and the farm need to be brought in line, in a dialectical way that embraces both adaptation and resistance, with a complex set of different but interdependent markets.

Together, these different markets constitute a constellation that is imposing what has been termed the squeeze on agriculture: upstream markets continue to impose price increases (thus contributing to cost increases), whilst downstream markets tend to offer lower or stagnant prices. Thus the available space between prices and costs is squeezed and the labour income goes down. Second, these different markets are increasingly world markets (and are less reflective of local, regional and/or national scarcity relations). Even if only some 16 percent of all agricultural produce physically crosses international borders, the presence and dynamics of food empires — extended networks that increasingly control the production, processing, distribution and consumption of food (Ploeg 2008) — imply that the same set of standards, parameters and procedures is applied at the global scale, thus also affecting all those products that are not traded and transported internationally. An important operational mechanism of food empires is that they increasingly delocate agrarian production and relocate it in areas where labour, land, water and environmental space are cheap and political support can be gained or bought. Alternatively they seek to shift production to areas with techno-institutional conditions that are favourable to large-scale corporate production. Such shifts and relocations can come as a sudden and complete shock to peasant farming. Access to markets is lost and complete regions can be wiped out economically. A third characteristic of the current market constellation is that it increases volatility. This is partly related to the previous points but also stems from speculation in futures markets. Finally, the markets for food and agricultural products are increasingly exposed to the consequences of the general economic and financial crisis. Credit to refinance existing arrangements is becoming scarce and/or more expensive, while the purchasing power of large groups of consumers is seriously affected.

All this implies that farms currently operate in a hostile and adverse context. The markets threaten, albeit to different degrees, the continuity of most, or nearly all, farms. This endangers employment levels, incomes and prospects for the future, while simultaneously bringing the possible destruction of a patrimony that has been built by generations. In short, the markets threaten to bring despair, misery and hunger, if they are not doing so already.

As indicated before, 1.4 billion people in the world live on incomes of less than $US1.25 a day (IFAD 2010). They live in extreme poverty. The majority (70 percent) live in the countryside. That is to say, there are one billion rural poor. Most of them depend partly, or to a large degree, on farming. Together with those who are only a little above this extreme poverty line, there are in total some three billion very poor people in the world. Many of them face hunger.

In large parts of Europe the majority of farmers earn less than the legal minimum wage level, while many face the threat of bankruptcy. In Eastern Europe particularly, the situation is dire (Bryden 2003)

Within this market constellation (an inevitable result of the current imperial food regime) the attempt to continue farming emerges as a form of resistance (Chayanov 1966: 267; Netting 1993: 329). Entering peasant farming anew is also an expression of resistance. It is not just a few who are engaged in resistance of this kind — it is a multitude. Many peasants are actively looking for and putting into practice adaptations, changes, new approaches and alternative patterns for cooperation. Thus, multiple processes of redesign are occurring that materially alter farm practices (for example enlarging multifunctionality and/or restoring autonomy). The same processes of redesign are also altering the ways in which farms relate to each other and to the wider context, leading to the emergence of new levels of resilience (as discussed by Oostindie 2013). This new resilience allows peasants to stay where they are, although discarded by major market forces and to prosper despite the tendency of external forces to spread misery and poverty.

From these flows of resistance, redesign and resilience new commons often emerge. This is the case with newly constructed market circuits in Europe, Brazil and China (Ploeg, Ye and Schneider 2012). Likewise, new commons emerge when peasant communities in Latin America regain control over their irrigation systems whilst simultaneously fighting the state or companies that try to appropriate their water rights (Boelens 2008; Vera Delgado 2011).

In chapter 6 I will discuss these new responses in more detail. Here it is important to note that the new and sometimes interlinked practices that result from redesign basically build upon the malleability of farming that I have discussed throughout this chapter in terms of Chayanov's thinking. In and through their daily struggles, today's peasants recalibrate several of the main balances that underlie the architecture of their farms and relink these balances in novel ways so that new styles of farming arise and mature, styles of farming that are at odds with the mechanics and needs of the surrounding systems. This results in the creation of new interstices that allow for, and require, further struggles and new, more encompassing responses.

By Way of Synthesis: The Peasant Farm

Building on the balances discussed so far (and several others which space does not allow me to discuss here)[8] it is now possible to present a synthesis of the peasant farm as it exists and functions today. This synthesis is meant to highlight three key issues. The first regards the relationships of today's peasant farms with those of the past: there is continuity as well as discontinuity and renewal (the latter two elements are partly due to the sharply changed politico-economic context). Second, this synthetic model embraces, as we will see later, both the South and the North: there is no fundamental difference, nor any intrinsic antagonism between peasants in different parts of the world. Third, the synthesis relates to marginalized peasant farms and poor peasant families as well as to highly productive and well-kept peasant farms and prosperous families. That is, it relates to reality and to the potentials contained in reality.

The peasant farm is the complex and dynamic outcome of the strategic deliberations and considerations of the farming family. Actual peasant farms, as they present themselves at a given moment in a specific space, are ever so many expressions of the art of farming that resides in the fine-tuning of each of the many balances entailed in the farm and in the skilful coordination of the different balances. Thus, fields and cattle are reshaped, plant varieties are carefully selected and improved, labour input is defined, capital is formed, knowledge is developed and networks are explored. The many balances are tied together in a coherent whole that translates into the organizational plan of the farm.

The art of farming, that is, the deliberate and strategically grounded construction of a farm and the many elements that constitute it, does not separate the farm from its politico-economic environment. Part of the art of carefully equilibrating many of the balances involves taking into account the parameters, opportunities and threats coming from this environment. These threats, opportunities and parameters are not translated in a straightforward linear way into the farm. They are, instead, always mediated by the farmer, who considers the different ups and downs. They are part of a balance that is equilibrated in a singular way by the farming family. Hence, general environmental tendencies will very often translate into differentiated effects. The art of farming is intrinsically interwoven with the reproduction of heterogeneity. The more so since the resulting heterogeneity becomes part and parcel of the deliberations: it provokes debates (which practices perform better?) and might induce changes (when ruptures occur the most resilient practices might inspire others and thus become a beacon for more extensive transitions).

The heterogeneity of peasant farms definitely does not "make any simple empirical generalization impossible" (Bernstein 2010a: 8). Considering the position of peasants in today's societies, and taking into account that their struggles for improved livelihoods mostly occur through the moulding and remoulding of their farms, we might, I think, very well elaborate six features that are both theoretically grounded and can be empirically validated.

The first and probably the most important feature is that peasant agriculture is geared to producing as much added value (or labour income) as possible under the given circumstances. Thus peasant agriculture intrinsically contributes to economic growth. There is one proviso, though. This contribution might become invisible. This occurs when the created value is appropriated by third parties, such as food empires or the state. This appropriation (or draining) might be so extensive that it slows down any further growth, capital formation and development in the countryside and even induces a deactivation of peasant agriculture (a form of involution that we are witnessing today).

The focus on the creation and enlargement of added value mirrors the peasant condition: facing a hostile environment through independently generating income in the short, medium and long term. In this respect the peasantry definitely is part of modernity, as recently argued by Lallau (2012) and Deléage (2012). Although the centrality of value added production within the framework of peasant agriculture might seem self-evident, it is a decisive feature in distinguishing peasant agriculture from other types of farming. The entrepreneurial mode of farming is as much oriented toward taking over the resource base of other farmers as it is to the direct creation of added value. Capitalist agriculture is centred on the production of profits, even if this implies a reduction in the total added value. Where conditions are equal for all three modes of farming, peasant farming emerges as the most productive one, realizing the highest yields and continuously working on further improvements of its own resource base. It also emerges as the most sustainable way of farming. All these statements apply equally to the developed and the developing zones of the world.

Evidently, the environment in which agriculture is embedded significantly influences the levels of value added and how they unfold over time. Peasant agriculture, in particular, requires space to fulfil its potentials. If such space is not available, due to the negative interactions of the environment on peasant agriculture, the ability of peasant farming to realize its potential is blocked. Thus, peasant struggles are a reflection of the multifaceted nature of the interactions between peasant agriculture and society at large.

A second feature regards the resource base available to each peasant unit of production and consumption: it is limited and nearly always under pressure (Janvry 2000). This is partly due to internal mechanics, such as inheritance practices that generally imply a distribution of limited available resources among a growing number of new households. It is also due to the external pressures on resources such as climatic change and/or usurpation of resources by large export oriented corporate interests. Generally peasants will not seek to counterbalance these pressures by expanding their resource base through establishing substantial and enduring dependency relations with markets for factors of production. Such strategies would run counter to their search for autonomy and would also involve high transaction costs. The relative and growing scarcity of available resources increases the importance of improving technical efficiency (see chapter 5). In peasant agriculture, this again implies achieving maximum output with the given resources, without compromising the quality of these resources.

A third feature relates to the quantitative composition of the resource base: labour will often be relatively abundant, while the objects of labour (land, animals, etc.) will be relatively scarce. In combination with the first characteristic, this implies that peasant production tends to be labour intensive, capital formation will often occur through labour investments and the development trajectory will be shaped as an ongoing process of labour driven intensification.

The qualitative nature of the interrelations within the resource base is also important. This points to a fourth feature: the resource base is not separated into opposed and contradictory elements (e.g., labour versus capital or manual versus mental labour), but, instead, the available social and material resources represent an organic unity that is owned and controlled by those directly involved in the labour process. In more political terms, it is a self-regulating unit. The rules governing the interrelations between the actors and defining their relations with the resources are typically derived from, and embedded in, local cultural repertoires, including gender relations. Chayanovian types of internal balances also play an important role.

A fifth feature (intimately interwoven with the previous ones) concerns the centrality of labour: the productivity and future development of a peasant farm critically depends upon the quantity and quality of labour. Associated aspects of this include the importance of labour investments (terraces, irrigation systems, buildings, improved and carefully selected cattle, etc.), the nature of applied technologies (skill oriented as opposed to mechanical) and peasant innovativeness.

In the sixth place, reference needs to be made to the specificity of the relations established between the peasant unit of production and the markets. Peasant agriculture is typically grounded upon (and simultaneously embraces) relatively autonomous, historically guaranteed, reproduction. Noncommodity flows and circuits are as important as commodity flows and circuits. Each cycle of production builds upon the resources produced and reproduced during previous cycles (see also figure 3.1). Thus, these resources enter the process of production as noncommodities that are used to produce commodities and at the same time help to reproduce the unit of production.[9]

The characteristics elaborated above flow together in the distinctive, albeit often misunderstood and materially distorted, nature of peasant agriculture, which is primarily oriented toward the search for and the creation of added value and productive employment. In the capitalist and entrepreneurial modes of farming, profits and levels of income can be increased through reducing labour input and/or taking over the resource bases of others (in whatever way). By contrast, peasant agriculture seeks to align the ongoing increase of added value per farm with increases in the value added by the peasant community as a whole.

At the level of the peasant community as a whole, the possession of a specific resource base by a specific family is generally recognized. Within the prevailing cultural repertoires (or moral economies), the takeover of adjacent plots or possessions is definitely not viewed as progress; for the peasant community as a whole this would be tantamount to self-destruction. Hence, individual peasant families strive to progress, albeit with different rhythms and different degrees of success, through their own efforts and using their own resources. This adds to the overall growth of value added at the level of the community or the regional economy. In capitalist and/or entrepreneurial farming, growth at the level of individual enterprises is typically associated with a stagnation or even decrease of the total amount of value added at higher levels of aggregation. A peasant economy excludes the occurrence of such a pattern.

A Final Note on Differentiation

In the preceding text some references have been made to another issue hotly debated by the radical left: the differentiation or stratification of peasant society. Heterogeneity in agriculture embraces many dimensions, but differences between smaller and larger farms (measured in whatever way) and poorer and richer families (often assumed to coincide with smaller and larger farms, although this is not necessarily the case) are often described in terms of the concept of stratification. This is based on the assumption that peasant society is composed of different strata — strata that are diverging (and developing into contrasting classes). Even so, many questions remain. What is the origin of the diverging trends that result in different strata? And what are the implications of stratification?

There are two contrasting views here. The Marxist/Leninist view centres on class differentiation. Opposed to this is the notion of demographic differentiation developed by Chayanov.

The canonical view on class differentiation is clearly specified by Marx (1951: 193–4).

> [the] peasant who produces with his own means of production will either gradually be transformed into a small capitalist who also exploits the labour of others, or he will suffer the loss of his means of production ... and be transformed into a wage worker. This is the tendency in the form of society in which the capitalist mode of production predominates.

In this scheme of things the countryside would eventually be populated by capitalist farmers, wage labourers who are working for them and peasant farms that have, as yet, not been dissolved. This last category then might be divided in three subcategories: "small" peasants, doomed to become proletarians; "medium-sized" peasants, who are "stuck in the middle;" and "large" peasants close to becoming capitalist farmers.[10] Chayanov's model of demographic differentiation provides a different view. He argues that differences in farm magnitude are basically temporary because they stem from changes in the consumer/worker ratio within the peasant family. A young couple starts with a small farm, but when the number of consumers grows in relation to the number of workers, the farm size will be augmented — until the couple grows old and the children go their own way; then the farm shrinks again. There are many variations on this theme as Chayanov (1966: 242–57) abundantly documents in his *Theory of the Peasant Economy*. Later, authors like Fei Xiao Tung (1939) showed that the demographic cycle might very well span four or five generations (see also Yang 1945: 132) and may imply considerable shifts in farming styles (Garstenauer et al. 2010).

Chayanov (1966: 248) was realistic in acknowledging that there were, in fact, "two powerful currents" in the Russian countryside of that time: class differentiation and demographic differentiation, the two often intertwined in complex ways. His position was later echoed by Daniel Little, who argued that both processes could occur, with the emphasis sometimes on the one, at other times on the other. The "Leninists," on the other hand, maintained that demographic differentiation, if it did exist, was irrelevant.

If we look back to the debates of that time with the benefit of knowing how history unfolded, we might state that, with a few exceptions, there has been no definitive class differentiation in worldwide agriculture since the 1880s that is as rigid and far-reaching as implied in the above quote. It has rather been the other way around. Especially during the international agrarian crises of the 1880s and 1930s, capitalist agriculture receded or even disappeared completely from large areas. This has been eloquently discussed and analyzed for the great American plains by Harriet Friedmann (1980 and 1993), and Zanden (1985) has documented the same phenomenon in Europe. Netting (1993: 296 passim) provides a general discussion of this phenomenon.

In the meantime the discussion has shifted to new mechanisms of differentiation — mechanisms that produce quite different effects from the ones expected more than a hundred years ago. A first new mechanism is related with the rise of entrepreneurial farming. This model works through takeovers, a strongly restricted or even taboo phenomenon in peasant agriculture. Agrarian entrepreneurs (a role model and identity that only emerged with modernization and the Green Revolution; see Ploeg 2003) take over land, water, quotas, symbols and market access from others, thus accelerating the process of quantitative growth at the level of the farm enterprise (see, for example, Gerritsen [2002], who documented this process for Mexico).

A second mechanism of differentiation relates to the current re-emergence of large capitalist farm enterprises, notably in the South (Schutter 2011). These have strong links to food empires or are even directly part of them. These new enterprises, currently also created through land and water grabbing, no longer compete with the peasant sector on prices. Their "competitiveness" is typically based on their control over channels (mostly global) through which agricultural products are bought and sold. Decisive in such control is privileged access, certification, standardization of products and volume of sales. It is, in short, "competitiveness" grounded on extra-economic coercion.

Together these new forms of differentiation represent very serious threats to the peasantries of today's world.

Notes

1. This was also expressed, more generally, by Chayanov (1923: 5) when he stated that "the biological nature of agricultural production distinguishes it from urban industry ... which is why the role of large and small enterprises in the former definitely differs from those of capitalist industry and artisanal units in the cities." This part of the introduction is missing in the Thorner edition. Mann and Dickinson (1978) subsequently developed this particular point of view.
2. This implies that current breeds of cattle, current plant varieties, and specific levels of soil fertility need to be understood as social constructions. They are the outcome of long and complex periods of co-evolution. See, for example, Sonneveld (2004).
3. This is an expression sometimes used by Chayanov. I will return to it further on in this chapter.
4. The social and intellectual power of these movements is explicitly recognized in the International Assessment of Agricultural Knowledge, Science and Technology for Development (IAASTD 2009).
5. The Italian agrarian historian Sereni (1981) beautifully described this process as involving *buoi rossi* [red oxen]: a metaphor for the controlled burning of pieces of woodland.
6. Within the neoinstitutional framework, "buying" involves transaction costs — costs above and beyond the price of the product purchased. For example, you may buy hay for a certain price. But if you don't know where it comes from (maybe a vineyard that was heavily sprayed with poison), there may be all kinds of risks (poisoned cows for example). This risk and/or the cost of getting information about the origin and quality of the product or service is called a transaction cost. From a neoclassical point of view there are no significant differences between the situation of actively constructed self-provisioning (i.e., relatively autonomous and historically guaranteed reproduction) and the one characterized by high market dependency. For a neoclassicist the

choice simply involves calculating the existing market prices. This is diametrically opposed to Chayanov's position (and that of institutional economists).

7. Here again, cultural repertoire (or moral economy) is crucial, especially the effect of more general values in curbing market opportunism. In this respect, Hobsbawm (1994: 342) refers to "fundamental motives of human behaviour" such as the "habit of labour." He argues that "the capitalist system, even when built on the operations of the market, had relied on a number of proclivities which had no intrinsic connection with that pursuit of the individual's advantage, which ... fuelled its engine" (ibid.). Aside from the habit of labour, such proclivities include

> the willingness of human beings to postpone immediate gratification for a long period, i.e., to save and invest for future rewards, pride in achievement, customs of mutual trust, and other attitudes which were not implicit in the rational maximisation of [profits]. (ibid.)

Hobsbawm argues that while capitalism partly relies on such values, it simultaneously destroys them.

8. These include the balance between the short and the long term that governs the interrelations between past, present and future; the balance between the known and the unknown; that between innovation and conservatism; and the balance between the peasant family and the rural neighbourhood or community. A wealth of information on these balances can be encountered in anthropological studies. See, for example, Durrenberger (1984) and Long (1984).

9. As discussed before, this pattern starkly contrasts with market dependent reproduction in which most or all of the resources are mobilized through markets, thus entering the production process as commodities. Then, commodity relations indeed penetrate into the heart of the labour and production processes.

10. This latter part is remarkably close to the classification schemes elaborated by neoclassical economics in the period 1960–2000.

4

The Position of Peasant Agriculture in the Wider Context

In the previous chapter I indicated that the different balances contained in the farming family and the farm unit have ramifications for broader and more general social relations, just as these relations are reflected within the farm and the family. While the internal balances are typically defined and set by the actors directly involved, this is not the case for these wider balances, which are discussed in this chapter.

External balances are not located within the family, the farm and/or between the two but at the interface between the agricultural sector as a whole and the society and markets in which it is embedded. These external balances cannot be set, or influenced, by individual farmers. These balances do, however, quite evidently have a considerable impact on individual farms and farm families.

Chayanov did not explicitly discuss or theorize about such external balances, although a part of his work can be interpreted as alluding to such influences (notably chapter 6 of the 1966 edition). There are clear references, for instance, to how the peasant economy might affect the labour market (Chayanov 1966: 240) — an issue that was much later taken up by Luiz Norder (2004) in Brazil. The same applies to state policies that affect the peasantry, as illustrated by Chayanov (1991) in the discussion on horizontal versus vertical cooperation in *The Theory of Peasant Co-operatives*. And then, of course, there is *The Journey of My Brother Alexis to the Land of Peasant Utopia* (Chayanov 1976), in which Chayanov, writing under a pseudonym, was able to express himself more clearly than in his other writings. This novel contains provocative discussions about the "optimal equilibrium between town and countryside" (Kerblay 1966: xlvii): in *Utopia* there are no longer any very large cities, and Chayanov also writes about agricultural intensification; the role of peasants in society; and provides us with prophetic flashbacks that (in 1920!) predict the end of Bolshevik rule and the establishment of direct democracy.

Town-Countryside Relations as Mediated by Exchange Relations

A first external balance concerns the interrelations between farms and downstream markets. Markets can operate differently at different times and in different places. Some will show a long run tendency of declining prices. Others will show, as Chayanov (1966: 105) remarked, "an improvement in the market situation" (see also ibid.: 83, figure 2.4). Italian farmers refer to such a situation as "*un mercato che tira,*" a market that pulls, i.e., a market that stimulates farmers to produce more. It is a market that allows for capital formation, since the prices received for the farms' produce are higher than the costs of production. Positive prospects (i.e., the expectation that prices will stay at a relatively high level) further contribute to such a situation. The opposite occurs when prices are low and expected to decrease further. Then we talk about adverse markets. These barely allow for the maintenance or reproduction of the farm; they inhibit further capital formation and hamper the farm's development. The producers have to endure such times, probably even have to considerably reduce their standard of living in order to survive. This market situation might emerge as the result of "urban bias" (Lipton 1977), global dependency relations (Galeano 1971) or from the squeeze that food empires currently impose upon agriculture.

These two market situations have different impacts on different farm types. The balance between internal and external resources on the farm can play a key role in how these forces impact at the micro level. Figure 4.1 summarizes these interactions in a simplified way. The arrow refers to the dominant trend in world agriculture over recent decades.

Figure 4.1 highlights the different equilibria that can exist between farms and markets. These equilibria are absorbed and translated by the farm units, affecting the different "internal" balances (utility, for instance, will be critically affected). However, the balance

Figure 4.1 The Interactions between Markets and Farms

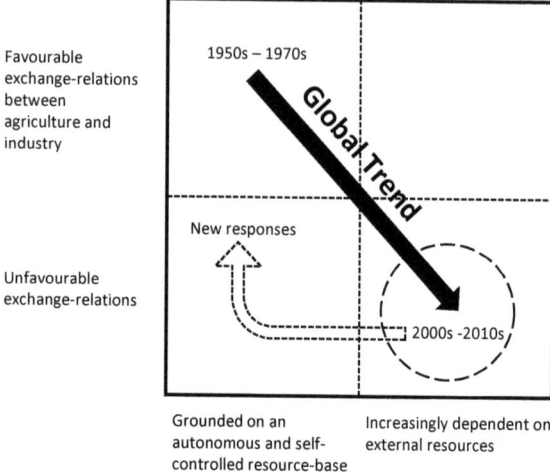

between farms and markets is not static. Peasants might retreat from specific markets and enter others (multiproduct farms have considerable flexibility here); they may use cooperatives as a countervailing power; when there are extreme misbalances they may rally in the streets calling for state interventions. They may even organize new market channels by themselves (Ploeg, Ye and Schneider 2012).

Many peasant holdings today are increasingly dependent on external resources and are simultaneously facing unfavourable exchange relations. Many of them became trapped in this difficult situation by the neoliberal project that has demolished agrarian policies, liberalized and globalized markets and unleashed all controls on capital. Neoliberalism has greatly contributed to moving agriculture from relatively autonomous units of production facing relatively favourable market conditions (this did not, of course, apply universally) to units that are strongly dependent on upstream markets (see previous chapter) and that face unfavourable market conditions (as illustrated by the arrow in figure 4.1). The consequence of such a move is that many farms, in both the South and the North, are finding it increasingly hard to continue.[1] In response, many farmers, all around the world, are seeking to move from the lower right position in figure 4.1

toward the lower left position in order to be better able to face the adverse markets: i.e., to make farming more peasant-like and more grounded on its own resources.[2] Some groups of peasants are even trying to move from the lower left position toward the upper left, through the construction of new markets and market channels. Both these attempts contribute to the richness and multidimensionality of today's peasant movements. Yet, the same attempts, while receiving much publicity, are still very much the exception rather than the rule.

Town-Countryside Relations as Mediated by Migration

Markets are not the only mechanism articulating agriculture and the urban economy — migration has been, and is, highly important as well. Migration can take many forms. It might be a one-way flow of people moving from the countryside toward the cities and the construction sites, factories, ports and informal sectors elsewhere. Extended slums in the periphery of the cities are the nearly unavoidable outcome of this process (Davis 2006). Rural poverty and/or warfare in the countryside may act as a push factor here, but the relatively higher wages sometimes paid in the urban economy (Chayanov 1966: 107) may also serve to pull people toward urban centres. Peasants often bring considerable skills to the urban economy. This was the case in Italy after World War II, when the *mezzadri* brought their networking capacities to the cities and created a blossoming sector of small and medium enterprises that became the heart of the Italian "miracle" (Bagnasco 1988).

The negative side, though, of the rural exodus, whatever its specific form, is the regression and abandonment that may occur in the countryside (Chayanov 1966: 107–8). Such negative effects can be avoided when migration is cyclical rather than unilinear, although other negative effects might emerge. Cyclical migration is characterized by youngsters leaving the countryside, experiencing city life, earning and saving money (usually only after getting engaged or married). Sooner or later these migrants return to their villages and invest in farming, shops and small enterprises. This pattern has often added considerable dynamism to agriculture. It has been important throughout Europe and is now important in China. It is impossible to understand Chinese agriculture without understanding the manifold cyclical patterns of migration that link it to towns and industries (Ploeg and Ye 2010). This cyclical pattern can sometimes also be transnational.

Historically, peasants working in the urban economy while maintaining their farms (often cared for by their wives or parents) have contributed to building a strong labour class, capable of standing firm in many conflicts. They could do so precisely because they had a fall-back position: their own farms. Ottar Brox (2006) has documented the example of Norway, where the labour class that emerged in the beginning of the twentieth century had rural roots and strongly contributed to decisive struggles that eventually resulted in a relatively fair distribution of the nation's socially produced wealth. This is still reflected in relations today. Norway is possibly the only oil producing country in the world where the huge benefits of this industry are used for the benefit of the population as a whole, rather than being grasped by oligarchies and private capital.

In short, migration is an important ingredient of the overall balance between town and countryside. Some forms of migration can sap the vitality of the countryside. Other patterns, instead, can strongly contribute to a revival of the countryside. One of the decisive factors is cultural repertoire: whether people judge returning to the countryside and improving the rural condition to be important or not.

Farming Versus the Processing and Marketing of Food

Historically, there has been an ongoing process of "externalizing" the processing and marketing of food. Today, most farming is limited to the production and delivery of raw materials, which are then processed by specialized food industries, many of which operate worldwide in an imperial way (Bonnano et al. 1994). Trade is increasingly controlled by large trading companies and retail chains. Alongside the agro-industries that control the flows of inputs into primary production, these industries, companies and chains compose networks (Vitali et al. 2011) that increasingly function as extractive systems.

4 / THE POSITION OF PEASANT AGRICULTURE IN THE WIDER CONTEXT

The interaction between primary producers and the food industry goes far beyond "simple" transactions of exchanging commodities for money. In his time, Chayanov (1966: 262) already observed that

> the trading machine, concerned about a standard quality in the commodity collected, begins to actively interfere in the organization of production, too. It lays down technical conditions, issues seed and fertilizers, determines the rotation, and turns its clients into technical executors of its designs and economic plan.

Later on, this aspect was thoroughly theorized by the Italian rural sociologist Bruno Benvenuti. He found the commodity relations to be accompanied by and intertwined with "technical-administrative" relations (Benvenuti et al. 1983). Together the two create an institutional framework that prescribes exactly what farmers need to do, when, how and in what sequence. This structure almost completely eliminates the "freedom to," as discussed in chapter 3. Consequently, the "agricultural entrepreneur" is, according to Benvenuti, a "ghost." Far from enjoying a wide margin of discretion for making entrepreneurial decisions, the agricultural entrepreneur is bound to a script defined by others, notably the food industry, trading companies, retail chains, input delivery industries, banks and state bodies (Benvenuti 1982; Benvenuti et al. 1988).

In Chayanov's day, cooperatives still offered the promise of an effective countervailing power. Cooperatives were class based (Chayanov 1991) and offered the peasant economy the advantages of large-scale operations:

> peasant co-operatives ... represent, in a highly perfected form, a variation of the peasant economy which enables the small-scale commodity producer to detach from his plan of organization those elements of the plan in which a large-scale form of production has undoubted advantages over production on a small scale — and to do so without sacrificing his individuality. He is able to organize them jointly with his

neighbours so as to attain this large-scale form of production. (ibid.: 17–18)

Today the situation is very different. Former cooperatives have evolved into entities that treat peasants in the same way as food empires. Consequently, new cooperative structures are no longer seen as offering promising linkages to the general commodity markets. Instead, new rural movements try to create new "commons": new markets embedded in new normative frameworks shared by producers and consumers. These new markets mostly emerge at the interstices — places where the functioning of large commodity markets is far from satisfactory. Equally, in food processing the terms of trade are no longer the primary issue to be negotiated; the main issue now is whether, and under what conditions, processing can be reintegrated in farming or the local economy. This question is particularly relevant since new, miniaturized technologies have the potential to make this a reality. Relocating processing and trading within the farm has become one of the key rallying cries of today's rural movements (Schneider and Niederle 2010).

State-Peasantry Relations

The state is an entity that reflects and governs — directly or indirectly — the relations between the urban and the rural economies and therefore the relations between markets and primary producers; the nature of migration; and the interrelations between peasants, traders and food processors. But it is more than this. The state is also an autonomous force that imposes its own imprint on rural dynamics. Thus the balance of power relations — the correlation of contrasting social forces — is a crucial feature that needs considering. Figure 4.2 illustrates this. It shows the ups and downs in yield levels (i.e., the physical productivity per unit of land) in an agricultural cooperative in the north of Peru. The yield level shown is based on the average yields of rice, sorghum, cotton, maize and bananas, all grown in the cooperative. The average is expressed as an index with the 1973–74 yields being equal to 100.

4 / THE POSITION OF PEASANT AGRICULTURE IN THE WIDER CONTEXT

Figure 4.2 Development of Yields in Luchadores, a Cooperative in the North of Peru, 1960s–1980s

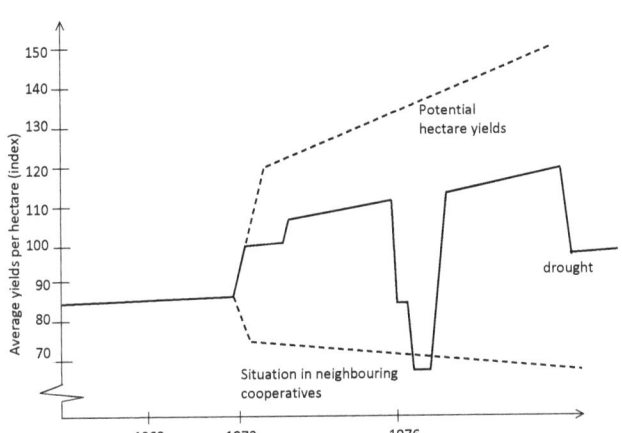

The key issue here is that yield levels reflect — almost with pinpoint accuracy — the power relations in the countryside, as mediated by the state. In 1969 a Land Reform Law was declared, but it was only in 1972, when the labour union decided to invade the lands of the large landholder and construct a new cooperative, that yield levels jumped considerably and kept growing. This neatly reflected the primary producers' increased power over the production process. This was the case until 1976, when the state intervened in the cooperative, took over the managerial reins and reduced employment by half. This caused a dramatic fall in yields that only recovered after a long-lasting strike and the withdrawal of the state appointed engineers. Then yields continued to grow until a dramatic drought struck the area in 1983.

The yields in this cooperative (*Luchadores del 2 de Enero*) were far higher than those of the neighbouring cooperatives. This was due to the presence of the labour union, which was translating the struggle for more employment into collective forms of labour driven intensification (see next chapter). As a matter of fact, the yields could have been even higher. The lack of sufficient "freedom from" (e.g., from the banking circuits, large trading companies and state agencies) was responsible for this relative underperformance (for a more detailed discussion see Ploeg 1990, chapter 4).

The balance between the state and the peasantry is, time and again, of enormous importance.[3] It often translates, as the previous example shows, into the fields and the processes of production located in them. The two sides of the balance have been masterfully described by James Scott. On one side "seeing like a state" (Scott 1998) is paramount; on the other, peasants excel in "the art of not being governed" (Scott 2009).

The equilibria that make up this balance often crystallize into specific agrarian policies. Many aspects of these policies have been criticized by the radical left. And indeed, they often run counter to peasants' interests (typically 80 percent of EU subsidies go to the richest 20 percent of farmers, notably the "agricultural entrepreneurs"). However, the baby should not be thrown out with the bathwater. Agrarian policies have been constructed, especially in the 1930s, to address and remedy profound and extended crises. This is true of the New Deal in the U.S. and the different agrarian policies in Europe that were later tied together into the Common Agricultural Policy. There is a continued and urgent need for agrarian policies in order to address the fundamental disequilibria in the relations between agriculture on the one hand and society, ecology and the interests and prospects of those directly involved in farming on the other. Designing policies that can reconcile these often conflicting interests is an urgent and challenging task. Establishing policies that promote equity and equality, or at least do not exacerbate existing inequities and inequalities, is particularly problematic, since agriculture at all levels is already characterized by major inequalities. For Chayanov (1988: 142) the "democratization of income distribution" was one of the major objectives of agrarian reform. However, at the global level there are deep chasms separating the North and the South (see Mazoyer and Roudart 2006), and such differences are also highly visible at the regional and local levels. As a result, agrarian policies almost inevitably have a highly differentiated impact, enriching some without providing sufficient assistance to those who need it. The costs and benefits of agrarian policies are often unequally distributed. It is, as yet, not very clear how this major issue can be tackled — especially when agrarian reforms are pushed to the margins of political agendas (Thiesenhuisen 1995). It is further complicated, by the peasantry's generally poor track record in resolving internal inequalities.

The Balance of Agrarian Growth and Demographic Growth

In assessing the balance between labour and consumption at the micro level (see chapter 2), the peasant farmer arrives at a required equilibrium between production and consumption. At the macro level this is reflected in the balance between agrarian growth and demographic growth as Ester Boserup (1970) showed to be the case for Africa. Demographic growth means that there are more mouths to be fed but also more hands to work the land. Thus it can induce agrarian growth. Similar relations have been described for other parts of the world. Huang (1990: 11) points to the centrality of demographic growth in the densely populated areas of China: "Population increase, acting through the distinctive properties of the peasant family farm, was what drove commercialization in the Ming-Qing Yangzi Delta, even as it was itself made possible by commercialization." Huang also recognized the other side of the equation: "the degree to which a peasant economy will involute depends very much on the relative balance between its population and available resources" (ibid.). Agrarian growth is subject to definite limits.

Today, in many parts of the world the once evident balance between demographic growth and agrarian growth is in disarray (see Netting 1993: 272). This is most visible, and dramatic, in Africa: the agricultural production per capita has been declining persistently for at least fifty years (Li et al. 2012). The once self-evident connection between production and consumption has been broken. This not only occurs at the level of nation states (thus triggering the call for food sovereignty), it also occurs at the micro level. It results in the tragic situation summarized in a Peruvian saying: *"tierra sin brazos y brazos sin tierra"* (land without hands to work it and hands without land). That is the typical situation of a rural household suffering from poverty and even hunger, while the land surrounding the homestead remains uncultivated. They lack the means to cultivate the land, and, for the moment, any possibility to readjust the completely distorted balance is beyond their reach.

Notes

1. The imminent danger linked to this tendency is that it induces strong downward trends in worldwide food production.
2. Here Mottura (1988: 27) observes that

    in periods with favourable agricultural prices the behaviour of the two groups of farms [represented in the left and right hand columns in figure 4.1] might be similar. However, as Chayanov indicated, the difference emerges in periods with bad prices. Then the latter group tends to slow down its economic activity, whilst the first one continues with the search for new opportunities in order to invest their labour.

3. Little (1989) argues that the power balance was decisive for the patterns of development in the countryside and indirectly for those in the cities as well. "In areas where peasants had been substantially deprived of tradition, organization, and power of resistance," other classes, such as "an enlightened gentry and budding bourgeoisie, were able through capitalist agriculture to restructure agrarian relations in the direction of profit and scientific innovation" (Little 1989: 119). However, in those regions "where peasant communities were able to defend traditional arrangements ... they could block the emergence of the property relations within which capitalist agriculture [and] wage relations in the countryside ... could emerge" (ibid.). See also Moore (1966).

5

Yields

The history of peasant farming is the history of ongoing intensification (see box 5.1). Over the centuries, farmers, both deliberately and unintentionally, have introduced small and sometimes larger changes in their production processes resulting in steady increases in yields. This process has been extensively documented by, amongst others, Slicher van Bath (1960), Boserup (1970), Wit (1992), Richards (1985), Bieleman (1992), Osti (1991), Mazoyer and Roudart (2006), Wartena (2006), Steenhuijsen Piters (1995) and Zanden (1985).

Yields are not merely technical parameters. They also reflect the complex and intriguing interplays between the micro and macro levels, between the local and the global. In other words, yields reflect social relations as much as they depend on them. Yields are the outcome of the labour process and thus reflect the ongoing adjustments in the many balances that order this process, particularly the balance between autonomy and dependence. Stagnating yields can result in abject misery or starvation; yield increases are the harbinger of more prosperous times and the prospect of greater emancipation of the peasantry. Higher yields also mean that agriculture can meet growing demands for food and nonfood products. Thus, at the macro level, yields are related to national balances of imports and exports and, more pertinently, with the strategic issue of food security.

Thorner's edition of *The Theory of Peasant Economy*, the most widely distributed and best known of Chayanov's works (at least in the Anglo Saxon world), pays hardly any attention to yields and intensification — both are only referred to in passing (see, for example, Chayanov 1966: 241). This reflects the Russian situation that was documented in the *zemstov* statistics. At that time, the end of the nineteenth century and beginning of the twentieth, there was no land scarcity in Russia, the more so since peasant communities regularly redistributed land. Consequently, peasant families' attempts to increase production and income occurred through expanding the amount of land they cultivated. However, in other publications, such as *Essays About the Functioning of the Peasant Farm*[1] (1924), Chayanov discusses the process of intensification at considerable length. It is a pity that this work is barely known beyond Italy (it was republished in Italian by Sperotto in 1988): it is fundamental for understanding today's peasant economies and, especially, for understanding labour driven intensification as an expression of peasant's struggles.

Intensification is the process that produces yield increases. It is "cultivating two spikes wherever there is just one spike now" (Chayanov 1988: 115). In his essays Chayanov equates peasant agriculture to high yield levels, observing a clear difference between the intensity levels of capitalist farm enterprises and peasant agriculture: "the level of intensity of capitalist agriculture is far inferior to the one of peasant agriculture" (ibid.: 117). This is due to three mechanisms. First, peasant agriculture goes where capitalist enterprises do not enter, opening up marginal lands and developing them into arable land or pastures. For capitalist enterprises, upgrading marginal lands is generally unprofitable (this, of course, depends on the average rate of profit in the overall capitalist economy). For peasants it is often a mechanism that allows for access to land, land that is constructed through peasant labour (ibid.: 80).

Second, peasant farms exhibit a far higher level of capital formation per unit of land (see chapter 2), using more seeds, more manure and more oxen or horses for traction per unit of land. "In the majority of cases, the farmer will increase the use of elements [such] as seeds, fertilizers, animals, etc., in order to produce more. Such increases will prevail over increases in the overall dimensions of the farm" (Chayanov 1988: 145). This is combined with a more intensive use of labour per unit of land, and together this allows for higher yields: "The better the land is worked (deeper and with more precision), the more it has been fertilized and the better the crops are cared for, the more intensive the farm will be" (Chayanov 1988: 146).

Third, the rationale governing the organization of production is radically different. A capitalist farm seeks to maximize profit, i.e., the difference between the gross value of production and costs, including labour costs. In the peasant farm the goal is to maximize the net product or labour income: the difference between gross production value and the costs of inputs, excluding labour (ibid.: 122). It is easy to demonstrate that this translates into different levels of intensity (see below). In short, peasants make improvements by converting idle land into a productive resource, combining it with higher levels of labour and capital and orienting production toward the highest achievable intensity. However, they can only do these things when they have the necessary politico-economic space (Halamska 2004).

Producing yield increases is far from being an element of secondary importance. For Chayanov (1988: 141) the increase of yields was part of the "development of productive forces" — to be considered explicitly as "a progressive phenomenon." Increases in yields might require "new relations of production" (ibid.: 142). By the same token adverse social relations of production might easily inhibit intensification or even provoke the opposite: extensification.

There is a nice detail in all this, a detail that is central to understanding the recurrent and divisive debates that have occurred about the "inverse relationship." The inverse relation is about small farms often having higher intensity levels than large farms. The empirical truth of this, its causes (provided it is true) and its implications (e.g., would dividing a large holding into smaller ones generate a leap in overall production?) are all hotly contested issues (see Sender and Johnson [2004] and Woodhouse [2010] for recent examples). For Chayanov this would be so much hot air. It is not about the difference between small and large. How can a small piece of land or a small unit of production by itself produce more than a larger piece or unit? Small or large units do not have intrinsic properties. While peasant farms are mostly (though not necessarily) smaller than capitalist ones, the essential difference is not one of size — it lies in the different modes of production. The peasant mode of production tends toward higher levels of intensity than the capitalist one, precisely "because there are radical differences between the objectives of capitalist farms and peasant farms" (Chayanov 1988: 72).

Intensification can basically follow two different trajectories: it can be driven by either labour or technology. Peasant agriculture is typified by labour driven intensification. The opposite trajectory is technology driven intensification, where yield increases are essentially the result of the application of new technologies and associated inputs. One could theoretically argue that the two are not incompatible and could be married together. However, in real life and within existing socio-economic relations, they tend to be mutually exclusive (see, for example, Hebinck 1990: 200). This does not mean that there is no technology in labour driven intensification or no labour in technology driven intensification. But the two involve designing and applying sharply differing techniques. I will come back to these crucial differences later in this chapter.

Box 5.1 Basic Concepts

All labour processes, including those within agriculture, involve three interacting sets of elements. These are the labour force, the objects of labour and the instruments. The labour process converts the objects of labour into products that contain more value — and often a different kind of value — than they originally had. One specific characteristic of agriculture is that the objects of labour are part of living nature. This is the case, for example, for fertile land that contains a rich soil biology that can deliver the nutrients required for plant growth. Land is always part and parcel of a wider ecosystem. Animals (that supply milk, meat, traction and manure), plants, fruit trees, vineyards, etc., are other objects of labour that all clearly represent living nature. The same applies to water, which people from Andean peasant communities perceive "as a sacred living being" (Vera Delgado 2011: 188).

The centrality of living nature strongly affects the agricultural processes of labour and production. It introduces variability and a certain unpredictability and requires permanent cycles of observation, interpretation, adaptation and evaluation. These activities are part of the artisanal labour process, the unfolding of which generates new insights that are decisive for the farm's production and reproduction (Sennett 2008).

The needed labour force might take many forms: men, women, children, neighbours who help each other. When participating in the process of production, they represent the labour force. The important point is that their labour converts the objects of labour into more useful items. This necessitates the use of instruments (or tools).

The instruments are used to facilitate and to improve the labour process. Just as with the objects of labour and the labour force, there can be an enormous diversity of instruments. Together with the knowledge carried by the labour force, the instruments compose a technique or technology. Here it is important to distinguish between skill oriented and mechanical technologies (see box 5.3).

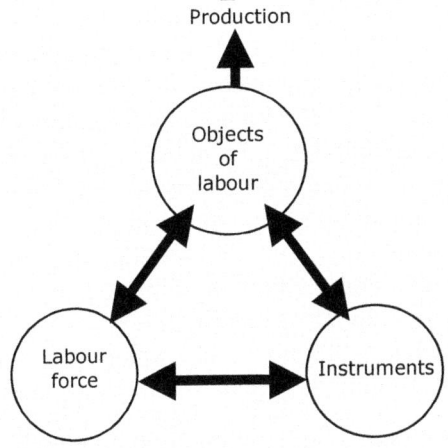

There are many possible combinations of labour force, instruments and the objects of labour. The nature of these combinations depends on the prevailing social relations of production. Such relations frame the labour process: they give it its concrete time and space specific form and dynamics. The social relations of production also govern the distribution of the wealth produced. These social relations consist of a wide range of factors, whose influence can vary greatly. Gender relations may be a key factor, or technologies, or relations between food industries and farmers and so on. When investigating the concrete patterns of farming it is always necessary to examine, within the specific empirical setting, the influential social relations of production. These relations are usually highly complex and continuously shifting patterns, made up of several interacting subsets.

The amount of value produced per object of labour (in agriculture also referred to as the yield) is understood, in agriculture, as the level of intensity. The higher the production per object of labour (e.g., the amount of grain produced per hectare or of milk per cow), the higher the intensity. Intensification refers to both increases in yields and to the process through which such increases are achieved. There are many different, and often highly contrasting, ways to intensify. The choices involved are hotly debated, and I will return to these later in this chapter.

Alongside intensity, the scale of farming is another key concept. This refers to the quantitative relation between the number of labour objects and the labour force needed to convert these objects into useful products (e.g., the number of hectares per worker or the number of milking cows per worker). The scale of farming depends on the instruments used and, more generally, on the social relations of production.

The interrelation between scale and intensity (see also chapters 2 and 4) is another much debated issue in peasant studies. These criteria are often used to assess and compare peasant agriculture with large scale corporate farming, which is often assumed to be superior.

There are different developmental trajectories in agriculture. Agriculture might develop through ongoing intensification. Or it might follow a different pattern, one of scale enlargement. And, of course, all kind of intermediate forms are possible. Hayami and Ruttan (1985) documented the different trajectories that can be observed internationally. They explained the patterns as reflecting relative factor prices (i.e., the relative prices of land and labour). If land is cheap and labour expensive, scale enlargement would dominate (and vice versa). This explanation has been seriously contested.

5 / YIELDS

Current Mechanisms of Labour Driven Intensification

Current forms of labour driven intensification are rooted in five mostly interdependent mechanisms. The first one, already identified by Chayanov as we saw in chapter 2, centres on the utilization of more labour and more capital per object of labour (see box 5.1 for this and other concepts). More labour is used per hectare or per animal and more tools and inputs ("capital" in the Chayanovian sense) are applied. This might lead to changes in cropping schemes, cultivation methods and/or increased care for the animals.

> Tillage, cultivation, and even harvest methods may be changed in their labour and capital intensity. For example, the same potato crop can be grown by using 40 or 120 workdays, with a corresponding harvest; a desyatina of fallow may have 1,000 or 3,000 puds of dung spread on it, and so on. (Chayanov 1966: 147)

Labour and capital (again, understood in the Chayanovian way) are used here in a complementary fashion: the one is not used as a substitute for the other.

The second mechanism involves fine-tuning the agricultural process of production. From a strictly agronomic point of view, agricultural production is based and depends on a wide range of what are known as growth factors, such as the amount and composition of nutrients in the soil, their transportability, the capacity of roots to absorb them, the availability of water and its distribution over time. The cultivation of wheat, practised for millennia, involves more than two hundred such growth factors and more are emerging with the development of scientific knowledge. Mixed farms, with different crops and animals (and "second tier" interactions) involve thousands of growth factors.

Crucially, these growth factors do not stay constant over time, they have not simply been there since the beginning of time: they are constantly changing, individually and as a whole. This is because they are constantly being regulated, modified and coordinated through the labour process. The amount and composition of nutrients, for example, are modified through the work of the farmer. The transportability of nutrients depends on ploughing, and the availability of water is regulated by irrigation and drainage. In short, the "behaviour" of growth factors is the object of specific tasks that are part of the labour process.[2]

Yield levels depend on the most limiting growth factor. Figure 5.1 shows the classic representation of these growth factors as the staves of a barrel. The yield, shown as the water level, depends on the shortest stave.[3]

Figure 5.1 Growth Factors and Yield Levels

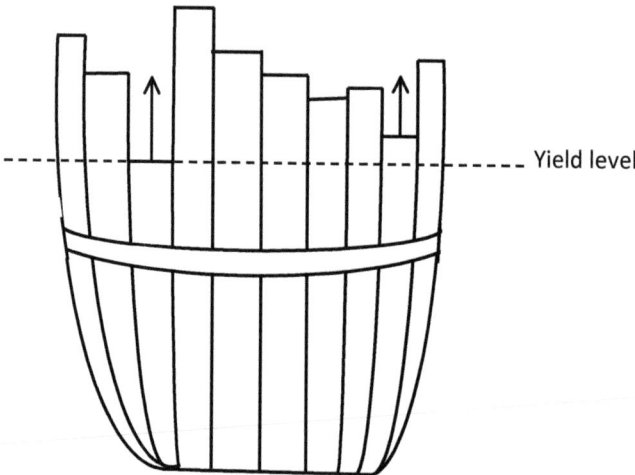

In their praxis farmers are continuously looking for the "shortest stave," that is for the limiting factor. Through complex and extended cycles of observation, interpretation, reorganization (often initially taking the form of experiments; see Sumberg and Okali [1997]) and evaluation, the limiting factor is identified and corrected. This leads to a change in existing routines, which, if successful, increase the yield level. This is an ongoing process: once the original limiting factor has been "enlarged," another will emerge as the new limit. The search for the shortest stave and its subsequent "rebuilding" is a process that generates knowledge. It is practical knowledge or *art de la localité*, as Mendras (1970) phrased it (see box 5.2). This type of knowledge unfolds through the process of labour driven intensification, which helps nurture and propel it, whilst simultaneously being enlarged by the resulting process. This particularly applies when conditions vary from place to place. *Art de la localité,* or local knowledge, is highly specific to time and place; it is artisanal and has a very different grammar from scientific knowledge (especially of the current technocratic type). It is knowledge that results in and is part of craftsmanship. The farmer is the carrier of this knowledge and craft. It is often knowledge *sans paroles*: experiential knowledge that is not (yet) articulated in unambiguous words. It is also closely associated with skills.

It is important to note that the search for improvements and the cycles of observation, interpretation, reorganization and evaluation are far from being individual enterprises. They often go beyond the single farm. They can involve extended networks for communication and sharing knowledge. They can cover considerable time periods and can also often assume a specific division of labour. These

Box 5.2 Local Knowledge

Cannon balls were fired long before military engineers understood the laws of ballistics. Ships sailed the oceans many centuries before Archimedes explained the law of upward pressure on a body immersed in fluid. Many practices are based on the skills of those involved and very often these practices are highly dynamic because the skills are continually being developed through the dialectical relation with the practices they inspire. Scientific knowledge *sensu strictu* is not always needed to generate new practices and/or improve existing ones. Often it is the other way around: scientific knowledge can be constructed because rich, heterogeneous and dynamic practices (of whatever nature) have already been developed. Science builds on such practices in order to derive and understand the laws entailed in them. From this it follows that science is not the only source of knowledge (although it is a very powerful source). Skill is another source, and local knowledge *(art de la localité)* can be an important part of this. Intuition can also play an important role.

networks are, as it were, the neural system of peasant agriculture, transmitting messages and receiving information from many different points. Sometimes such networks are converted into important mechanisms within socio-political struggles in the countryside (see, for example, Rosset et al. 2011).

The cycle that goes from observation to the evaluation of adaptations is crucially dependent on knowledge, just as it enlarges the available stock over knowledge. Here we are dealing with experiential, practical or local knowledge. Together, the ongoing process of fine-tuning and the resulting development of knowledge lead to a particular type of technology, which Francesca Bray (1986) referred to as skill oriented technology (see box 5.3).

From a technical point of view, successful fine-tuning increases the technical efficiency of the process of production in which the same amount of resources is used to realize an increased level of production. This increase of technical efficiency crucially depends on the quality of labour.

A third important mechanism in labour driven intensification resides in the systematic improvement of the resources used (Boelens 2008). A resource might be improved through a carefully calibrated balance of production and reproduction. This usually occurs in a step-by-step way, although sometimes considerable jumps occur — then

> **Box 5.3 The Contrast Between Mechanical and Skill Oriented Technologies**
>
> The Western worldview often associates technology with increased physical yields and technical efficiency. However, as Francesca Bray (1986) showed in a beautiful study on "rice economies," this is not necessarily the case. Bray distinguishes between mechanical and skill oriented technologies. Skill oriented technologies use relatively simple instruments (see box 5.1) combined with the skill and knowledge of those who work with them. With mechanical technologies it is the other way around: while the instruments are very sophisticated (for example, automated milking machines), they require very little knowledge to operate. Hence, mechanical technologies often bring about de-skilling.

there is an abrupt and substantial leap forward. Either way, there is a process of improving the fields (through manuring, terracing, constructing irrigation and drainage facilities, levelling, deep-ploughing, etc.); strengthening of soil biology (thus augmenting the capacity of the soil to generate nitrogen); improving breeds to make them more productive and better adapted to local circumstances (through processes of selection, crossing and culling that extend over long periods of time); constructing new buildings (to reduce harvest losses, for example); creating new varieties (through interplanting and spontaneous cross-breeding, testing and multiplying); enlarging local knowledge; developing skills; and unfolding new networks. In practice, such improvements often flow together with activities that belong to the first and second mechanisms (more labour and more capital per object of labour and fine-tuning, respectively). However we have to analyze this process separately. It is the third mechanism (improvements) that allows labour objects to absorb more labour and capital (i.e., the first mechanism). In its turn, the improvement of resources often follows in the aftermath of the second mechanism's cycles, which try to identify the shortest stave in the barrel.

A fourth mechanism, closely associated with the ones discussed so far but presented separately here, is novelty production. Novelties are

> located on the borderline that separates the known from the unknown. A novelty is something new: a new practice, a new insight, an unexpected but interesting result. It is a promising result, practice or insight. At the same time, novelties are, as yet, not fully understood. They are deviations from the rule. They do not correspond with knowledge accumulated so far. (Ploeg et al. 2004: 200)

To echo Rip and Kemp (1998), a novelty is, "a new configuration that promises to work."[4] Over the centuries farmers have achieved steady increases in yields through novelty production. This process has been amply documented: Ye (2002) gives an informative account of novelty production in China in the years after decollectivization (see also Ye et al. 2009); Osti (1991) and Milone (2004) documented novelty production in the periphery of European agriculture; Adey

(2007) does the same for southern Africa; and Wiskerke and Ploeg (2004) provide a general overview.

Novelties often remain hidden within local agricultural practices. Their dissemination can be slow and limited. However, novelties can also be identified and taken up by researchers who test and develop them further and eventually reintroduce them, in an improved and consolidated version, into the agricultural sector. Such flows (and the resultant cooperation between scientists and farmers) have proved to be highly powerful mechanisms. However, after World War II, when agrarian sciences began to follow a far more technologically driven path, they became the exception rather than the rule. Currently agroecology (Altieri 1990; Altieri et al. 2011) is leading the way in building on novelties and unfolding them into more widely applicable improvements.

Novelties can be incremental, building on each other and resulting in small, cumulative yield increases. Equally, they might be radical: introducing complete changes in existing practices and bodies of knowledge and producing abrupt and considerable jumps in yield levels. A current example of such a radical novelty seems to be represented by the system of rice intensification (SRI), "a set of practices and principles, rather than a technology, to be followed and implemented flexibly and in response to diverse agroecological and socio-economic conditions faced by farmers" (Stoop 2011: 445). It is telling that "SRI emerged in relative isolation from the international mainstream of rice agronomy" (Maat and Glover 2012: 132). SRI actually emerged out of cooperation between de Laulanié, a French priest with an agronomic background, and rice growers in Madagascar. It was born out of scarcity and adverse weather conditions. Each single step in the rice-growing practice there intuitively seems to be counterproductive. The practice involves planting very young seedlings, widely spacing individual tillers, alternating between wet and dry soil moisture regimes (instead of permanent flooding), using organic rather than mineral fertilizers and weeding frequently. However, together these changes have produced spectacular jumps in yields that are accompanied by considerable cost decreases, and together these factors explain the wide dissemination of SRI, which is now practised in many countries. In retrospect, SRI represents a paradigm shift: it is a definite move away from the model that views more plants per hectare and more fertilizer as the ways to achieve higher grain yields. In contrast to the varieties promoted by the Green Revolution, the cultivars used in SRI are built on their tillering features, with an emphasis on developing an abundant root system.[5] These better developed and more active root systems increase drought tolerance as well as efficiency in nutrient uptake and thereby reduce fertilizer use (Stoop 2011: 448). At the same time, building a healthy supply of soil organic matter strengthens the beneficial associations between roots and soil biota.

SRI is a radical, far-reaching, convincing and powerful change that has been created from praxis and outside the realm of institutionalized agrarian science. It was initially neglected, if not actively derided, by the scientific establishment. I will return to this point when discussing the "ghost" that seems to be one of the biggest limitations to labour driven intensification: the so-called "law of diminishing returns."

The fifth and final mechanism is the specific calculus used in peasant agriculture to optimize agricultural production (see box 2.4 and note the centrality of "good yields"). Peasants strive for the highest possible labour income, which differs significantly from the search for the highest possible profit on invested capital (Chayanov 1988: 73). In doing so, they drive the other four mechanisms (that carry intensification) as far as they can.

Building on the approach developed by Chayanov I will try to explain this fundamental point in two steps. The first step uses a simple function of production, as given in figure 5.2. It describes the physical input/output relations that characterize the production of, say, barley at a given moment in time. After more fine-tuning, or when some novelty has been created, the function might very well shift, but at this given moment it is as represented in figure 5.2. Let us assume that one unit of output renders one euro. The same applies to inputs: one unit costs one euro. The labour input (say in hours) is also given, below the x axis. Let us assume that one hour of labour (in the case of wage labour) also equals one euro. Total costs refer to the cost of inputs used plus the labour costs.

Figure 5:2 A Function of Production

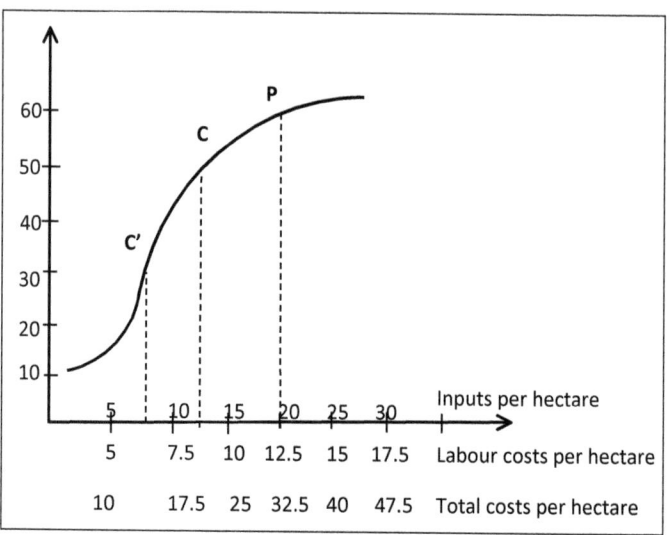

Now, if the production of barley was located in a peasant unit of production, the peasant would, if possible,[6] go to input level 20, which renders a production of EUR 58 (at point P on the production function). Why? Because going any further would make him the proverbial fool: going from input level 20 to 25 he would spend five extra euros but only make an extra four. In contrast: going from 15 to 20 costs five euros and renders six. Thus, at the input level of 20 (or a little beyond) he will get the highest possible labour income (output minus input). In this case his labour income will be EUR 38 (the difference between 58 and 20).

If the same crop were cultivated in a capitalist farm enterprise, things would be calculated differently. The entrepreneur would not be interested in maximizing labour income but in optimizing profit on invested capital. The highest profit (taken in isolation) emerges around input level 12 (this is at point C). Going toward that point implies that extra benefits are higher than total costs, which include inputs as well as wage labour; beyond this point the extra benefits are lower than the extra costs. At the optimum input level (12), the profit is EUR 27 (48 - 21). The highest return on investment (i.e., the highest profitability), however, emerges at lower levels of input and labour investment and consequently results in a lower level of production. Net profit as percentage of total costs is some 135 percent at an input level of 7.5 (this is at point C'); at input level 12 it is around 120 percent. This shows that, in a theoretical world, peasants achieve higher levels of intensity than capitalist farmers. The former produces at point P in figure 5.2, the latter at point C or C'. This is because the way of calculating is different. The peasant is interested in optimizing labour income (total production minus inputs). The capitalist farmer looks for the highest return (total production minus inputs and wages). The first equation moves the peasant toward point P, the second moves the entrepreneur toward point C'.

All this is, of course, terribly hypothetical. There are many reasons why the slope of the functions of production might be different for peasants and for entrepreneurs. There might be differentiated prices, or specific spending or agricultural policies and support systems that are more favourable to one group than to the other. The point, though, is that under equal conditions peasants produce at higher levels of intensity than capitalist farmers.

In real life, "equal conditions" are rarely found — especially in today's agriculture, where peasants operate alongside powerful capital groups. It is also important to note that peasants and capitalist entrepreneurs seldom use the same production methods. The latter increasingly have access to technologies that are beyond the reach of peasants. This might well blur the "inverse relationship," although this is not necessarily the case.

In the early 1980s I became familiar with rice production in the Peruvian Costa (the coastal area). At that time four technological levels could be distinguished. These are summarized in figure 5.3.

The first column illustrates the situation when the farmer transplants the seedlings rather than sowing the seeds directly in the field. This requires far more labour — although it saves labour when it comes to weeding — and results in the highest yields. Most of the inputs used (e.g., seeds, dung) are produced on the farm itself. This is a pattern frequently encountered in peasant farms. They do not perceive a high labour input to be a problem: high yields guarantee

Figure 5.3 Technologies, Yields and Cost Levels

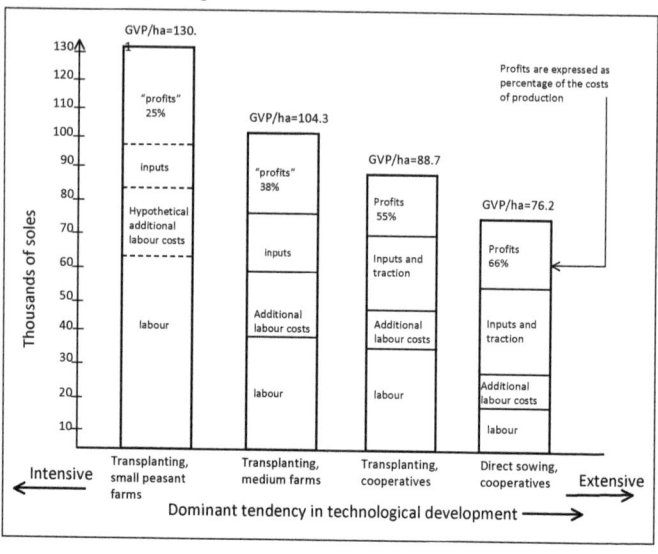

a good labour income. The second and third columns (encountered in medium-sized farms and cooperatives) combine the transplanting of seedlings with a higher use of inputs acquired through the markets (notably fertilizer and herbicides) and more mechanization (see the costs for traction in the third profile). Labour (especially in the third case) is wage labour.

The fourth column uses mechanized direct sowing (it can even be done with small aircraft). Wherever feasible the other tasks, such as crop protection and harvesting, are also mechanized. Yields are far lower, especially compared to peasant farms. However, profitability (the ratio of profits and costs) is the highest in this case even if profits are, in absolute terms, lower than in the second profile (66 percent versus 38 percent). Thus, the combination of the *banco agrario*'s unwillingness to finance high spending per hectare and a management that aims at a high return on investment (which also makes it highly risk averse) brings forward the lowest yields — paradoxically through applying the most "modern" technologies.

It was only after considerable socio-political struggles that workers in some cooperatives persuaded the management to introduce "the creation of productive employment" as a leading objective. This led some large cooperatives to turn to the first technological profile, thus "painting the fields green and helping Peru to feed itself," as my compadre Perez said at the time (see Ploeg 1990: 205–58).

The Significance and Reach of Labour Driven Intensification

The potential reach of the five mechanisms discussed above — and consequently, the potential of labour driven intensification — has often been neglected or grossly underestimated within peasant studies and other related disciplines, such as agrarian and development economics. One of the key concepts used in all these disciplines is the law of diminishing returns. This is grounded, basically, on marginalist logic, which assumes that as one adds more resources (for example, applying more labour per hectare), one obtains less and less extra production. At a certain point the relationship can even become negative. When applied to peasant society as a whole, these diminishing returns would translate into structural involution, the very opposite of development. At first glance the diminishing return argument sounds convincing. If too much seed is sown in a field the plants just dislodge each other, too much fertilizer can poison the soil and too much water will drown the plants. However, peasants don't want to be seen as the village idiot. They will refrain from overusing particular inputs and will instead look for the "shortest stave" and reorganize their farming practice so as to intensify without falling into the trap of diminishing returns.

In theoretical production ecology it is argued that diminishing returns are the exception rather than the rule (Wit 1992). In farming, diminishing returns trigger the search for new solutions and are the drivers of further progress (such as SRI). Then farming jumps toward a new function located at a higher level of productivity (see figure 5.4). And once a new solution has run against its own limits, the same basic procedure is repeated. Thus, an overall trajectory emerges that is characterized by increasing returns (as illustrated in figure 5.4). These increasing returns ultimately reach natural limits, basically associated with the availability of light and the upper limits of photosynthesis on which all plant growth is built (not to be confused with the limits related to sustainability). However, agriculture, wherever located, is a long way from reaching such natural limits.

Figure 5.4 Diminishing Returns as a Special Case, Not the Rule

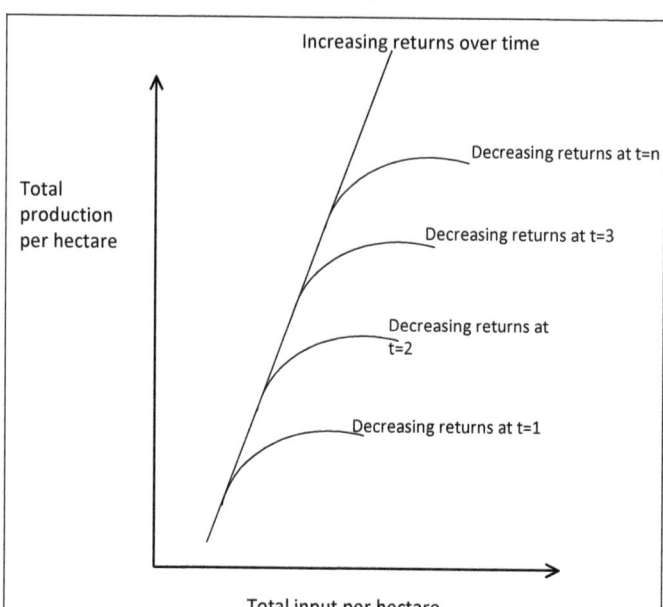

Ironically, Lenin was among the first to anticipate these insights from today's theoretical agronomy. Whilst Chayanov (1988: 88) wrote of the law of diminishing returns, Lenin (1961: 109; italics in the original) argued as early as 1906 that the law of diminishing returns was

> an empty abstraction, which ignores the most important thing — the level of technological development, the state of the productive forces. Indeed, the very term "additional investments of labour and capital" *presupposes* changes in the methods of production, reforms in technique ... [N]ew machinery must be *invented* and there must be new methods of land cultivation, stock breeding, transport of products,

and so on and so forth. Of course, "additional investments of labour and capital" may and do take place even when the technique of production has remained at the same level. In such cases, the "law of diminishing returns" is applicable *to a certain degree*, i.e., in the sense that the unchanged technique of production imposes relatively very narrow limits upon the investment of additional labour and capital. Consequently, instead of a universal law, we have an extremely relative "law" — so relative, indeed, that it cannot be called a "law," or even a cardinal specific feature of agriculture.

All this explains, according to Lenin, "why neither Marx nor the Marxists speak of this 'law,' and only representatives of bourgeois science ... make so much noise about it" (ibid.: 110).

Since these polemics took place, peasant agriculture has repeatedly shown itself capable of turning away from routes that lead to diminishing returns and of creating a trajectory that produces increased returns (see, for example, Richards [1985] on western Africa and, more generally, Netting 1993). Nonetheless, the field of rural studies is still haunted by the ghost of diminishing returns (see, for example, Warman 1976; Yingfeng Xu 1999; Barrett et al. 2001).

When Labour Driven Intensification is Blocked

The possibility of increasing returns (see figure 5.4) does not imply that stagnation, regression and even involution cannot occur. On the contrary. The point is that they are not intrinsic features of peasant agriculture. They become a feature of peasant agriculture as a result of particular politico-economic patterns and regimes.

Stagnation can occur for many reasons. It can be a result of highly unequal exchange relations. This situation precludes the peasantry from making any reassessment of the balance between utility and drudgery, simply because all the utility is appropriated by others. It can also occur because water is taken away (Vera Delgado 2011) or whenever peasant agriculture is locked into homelands as in South Africa during Apartheid or small rice producing "pockets" located alongside export oriented plantations as in colonial Indonesia (where

Clifford Geertz [1963] elaborated his theory on agrarian involution). And regression occurs wherever and whenever rural poverty is so high that the only hope the sons and daughters can see is to escape to cities, to carry bags or sell their bodies. Then there is nobody left to carry manure to the fields, care for the herd or maintain the dikes that surround the rice polders (as has occurred in Senegal, Gambia and Guinea Bissau). Regression also emerges in strongly patriarchal societies, where mothers tell their daughters to "marry whomever you want, as long as it isn't a peasant" (this happened, for instance, in large parts of Spain, areas that are now nearly completely deserted and desertified).

Peasant agriculture also regresses whenever new capital intensive technologies are applied in large-scale corporate farm enterprises, wherever located, thus allowing the latter to outcompete peasant farms producing the same goods and crowd them out of the market. This occurs especially when free trade arrangements prevail and environmental damage is not taken into account.

These forms of involution, stagnation and/or regression are all expressions of the agrarian question. We talk of the agrarian question when the relations between the way of farming (the concrete organization of the agricultural sector) on the one hand and society, ecology and the interests and prospects of those directly involved in agriculture on the other are out of balance. In the examples discussed above the peasants experience poverty, whilst society does not receive the extra food it needs (which may also harm the process of capital accumulation). In 1917 Chayanov dedicated an important essay to the agrarian question "*Čto takoe agrarnij vopros?*" ("What, then, is the agrarian question?") in which he links the emergence of an agrarian question to the way the social relations of production are organized (Chayanov 1988: 131–72). This leads to another important conclusion: that agrarian reform necessarily implies a thorough reordering of those relations. It can never be reduced to a simple distribution of land (it is telling that Chayanov rejected the populist slogan, "land to the tiller," that would later play such an important role in Latin America). Agrarian reforms need to aim at a "maximum labour productivity in agriculture," a "democratic redistribution of the national income" (presumably implying a correction of biased town-countryside relations) (ibid.: 142) and, finally, need to avoid "that any *desjatína* remains without sowing, or that any herd will be abandoned or slaughtered" (ibid.: 158). An agrarian reform implies a socialization of the land (ibid.: 156), which cannot be realized through a kind of "enlightened absolutism" (a sharp critique, *avant la lettre*, of Leninism and Stalinism) but is to "result from the involvement of local and democratically elected councils" (ibid.: 164). "Only then can sufficient contributions to the building and development of the nation state be delivered" (ibid.: 172).

What Propels Labour Driven Intensification?

The answer to this question is simple. Intensification is driven by the peasantry's search to improve income or, more specifically, their pursuit of extra added value in order to improve their labour income (Hayami 1978; see also chapter 2). Whenever and wherever the peasantry aspires to further improvements, and these aspirations are not inhibited by unfavourable social relations, this translates into and occurs through an increase in production. This is one of the fundamental issues brought forward by Chayanov. He also demonstrated this interrelation empirically (see, for example, Chayanov 1966: 99 and especially table 3.13). If there are more "workers in the family" (as regards the balance of labour and consumption) and if there is more "fixed capital per worker" (as regards capital formation and the balance of utility and drudgery on which it builds), then the "total family income" increases as well. This is because "more workers in the family" and "more capital per worker" translate into an increase of "area sown per consumer" and therefore in increased production (if no extra land is available, this will translate into intensification to achieve increased yields). In short, increases in food production link the emancipation of the peasantry to the progress of humanity as a whole — and it is precisely this link that has shaped agrarian history.

Today, just as in the past, there are many situations in which the (labour) income of peasant families may be under considerable stress. This might be for a variety of reasons: a price cost squeeze, lack of access to markets, heavy taxation or many others. In such situations, the search for improved income becomes part of a multifaceted social struggle. "[T]he farm family uses, within its power, all the opportunities of its natural and historical position and of the market situation in which it exists" (Chayanov 1966: 120). When external pressures threaten the continuity of the family farm, the search for more added value is part of a more general resistance.

Intensification and the Role of Agrarian Sciences

There are two basic narratives that can be used to explain the interrelations between agrarian sciences and agrarian growth. The hegemonic storyline is that the dynamics of agriculture (and notably the ongoing increases in productivity) are essentially due to a constant flow of innovations that go from science and enter the practice of farming. This storyline strongly reduces the role of peasants themselves, if not completely ignoring any role they might play. Telling examples are found within the many studies that try to assess the benefit/cost ratio of agricultural research. These studies simply take all productivity increases in agriculture as "benefits" and relate them to the "costs" incurred for agricultural research and technology development. Peasants themselves are absent from this picture and the results of their efforts are attributed exclusively to agrarian sciences.

We can also discern a second narrative, diametrically opposed to this first one. It is a less developed and more embryonic narrative and does not attract support from agricultural universities, agro-industries, ministries of agriculture and other institutions. Despite this, its roots and expressions can be encountered in many places. Chayanov's *Social Agronomy* (1924) is one important expression of it. In sketching his social agronomy, Chayanov built on the work of agronomists such as the Italian Bizzozzero, who were deeply involved in farming practice. Social agronomy soon became a point of reference for others, notably in Europe, at least until the outbreak of World War II. After the war, the hegemony of U.S. agricultural sciences meant that social agronomy disappeared, both as practice and as a point of reference. Only in more recent times, are these practices being revived and their relevance being rediscovered.

This second narrative basically argues that most agricultural renewal stems from farming practices. Instead of being the final destination of innovations, it sees the farm as their main origin. Novel production methods generate new insights, practices, artifacts and techniques. Some of these are picked up by research institutions, which further develop and disseminate them. This might be a "friendly" process, improving novelties so that they can be circulated on a wider scale. It can equally be a "hostile" takeover, selecting and appropriating a few that can be rebuilt and patented in a way that serves the interests of actors other than the originators and suppressing or ignoring those novelties that cannot be appropriated.

Many studies support this storyline of the peasantry being a main producer of novelties. Such narratives emphasize the interplay between peasants and research institutes. Paul Engel (1997) examined the sources of innovative ideas communicated by Dutch extensionists to Dutch farmers. He found that 40 percent of these ideas came directly from novel practices developed by pioneering farmers. Another 40 percent came from other extensionists, who themselves obtained most of their new ideas from farmers. Only 20 percent was directly derived from research stations and the like.

Vijverberg (1996) studied the dynamics of horticultural research in the Netherlands. He distinguished between innovations originally proposed by or taken from horticultural growers and others proposed by researchers or derived from science generally and/or other economic sectors. Those in the former category resulted in positive and widespread dissemination, whilst the latter type frequently failed. Too often there was a mismatch with practice: the new techniques and/or artifacts did not match with the horizon of relevance of horticultural growers; they did not suit the conditions within which growers operate, their interests and prospects and the specific way in which their labour process is structured.

Vijverberg's findings are echoed, in more general terms, by Mazoyer and Roudart (2006: 398) who concluded, after reviewing different historical trajectories of change in agriculture, that

> no machine, no product, no procedure can be designed and developed without calling on the acquired experience and the ac-

tive participation of technicians and practitioners themselves. The proper functioning of the chain of innovations requires that researchers, teachers and students at all levels know the practice intimately, its constraints and its needs. Otherwise, many new inventions end up being inadequate, are rejected, and become an incredible waste of resources.

Despite such historical lessons, the experiences, views, interests and prospects of the practitioners are all too often neglected. This occurs especially when agro-industrial interests become the framework that specifies the architecture of innovations. This can lead to the aborting of promising alternatives and can strongly distort agrarian growth and development.

Regardless of the many negative experiences and notwithstanding the promising alternatives, agrarian sciences that operate in splendid isolation continue to occupy a central position in hegemonic discourse (and to claim the lion's share of available resources). One of the pillars of this hegemony is the claim that only science and capital will be able to feed the world in 2050 — a claim to which I will return at the end of this chapter. There are three other factors that seem to lend strong support to the narrative that maintains the central position of agrarian sciences. These are the invention of chemical fertilizers (and the associated "chemicalization of agriculture" [Mazoyer and Roudart 2006: 376]), the mechanization of farming and the development of high yielding varieties.[7] All three apparently produced major and long-lasting leaps in agricultural productivity breakthroughs that, it is generally thought, farmers could never have created by themselves. These three examples are used as a testimony to the enormous power and potential of agrarian sciences.

In relation to the first factor, it is important to recognize that farmers have been improving soil fertility over the ages — long before von Liebig discovered the principles that govern chemical fertilization. "Agriculture that excludes fallowing (because it actively reproduces soil fertility) had been practised since the fifteenth century in Flanders, Brabant and Artois, without being the creation of any agronomists" (Mazoyer and Roudart 2006: 347). According to Chambers and Mingay (1966: 2), Great Britain's Agrarian Revolution

(1750–1880), which saw agriculture increase its output and feed 6.5 million more people in 1801 than a century before, "was not, to any significant extent the result of [exogenous] innovation."

> It is not on grounds of technological innovation that English agriculture can be said to have experienced a revolution. Except for an eddy here and there, the "wave of gadgets" that is said to have swept over England passed it until well into the nineteenth century ... As early as 1800 ... British farmers and landlords had accomplished the feat of releasing the latent powers of the soil on a scale that was new in human history. (ibid.: 3)

This began

> with the development of convertible agriculture, involving the alternation of arable and grass in place of the ancient division of the cultivated area between permanent arable and permanent grass, which tended to undermine the fertility of both. Alternate agriculture implied the practice of arable farming for fodder crops, i.e., the laying down to grass of parts of the arable in temporary leys and the sowing of legumes such as clover, sanfoin or lucerne which added to soil fertility while yielding heavy crops of hay. Hence, subsequent arable crops had the double advantage of an increased supply of animal manure and an enhancement of natural fertility through the [nitrogen fixation] of the fodder plants. When, in the second half of the seventeenth century, turnip began to be cultivated as a regular field crop, calling for heavy manuring and meticulous weeding, the foundation was laid for a new form of land use, especially adapted to light soils, hitherto sustainable only for rough grazing. (ibid.: 4)

The range of methods available for the reproduction and increase of soil fertility was continually enlarged. For example, the introduction of guano (seabird excrement accumulated along the Peruvian and Chilean coastline) played an important role. This was followed by the introduction of chemical fertilizers, notably after

World War I, when the industries producing explosives switched to the production of chemical fertilizer. Initially, peasants used their knowledge to combine the use of chemical fertilizer with other available methods: the use of "well-bred" manure and cultivation techniques that stimulated the soil's own capacity to generate nitrogen. Only far later did increased dosages of chemical fertilizers start to suppress the positive contributions that these methods made to maintaining soil fertility: the careful production of manure was increasingly abandoned (it required too much labour and no longer fit with the ruthless search for scale increases), and animal dung became waste and new technologies were designed to get rid of it as quickly as possible (later, less polluting and more environmentally friendly mechanisms had to be developed).

In synthesis, it is deceiving to present chemical fertilizer as heralding the absolute superiority of agricultural sciences over farmers' knowledge systems. The true story is different: it is the tragedy of losing a potentially highly valuable resource (manure), a loss to which agrarian sciences substantially contributed by paving the road for chemical fertilizers only. Chemical fertilizers became powerful, convincing and indispensable because manure, soil biology, mixed cropping as in the Central American *milpa*, complementary intercropping, green fertilizers like clover and local repertoires for making "well-bred" manure were neglected and, in the end, even treated as "monstrosities."

The "motorization" of agriculture (a term used by Mazoyer and

Box 5.4 The Contribution of Justus von Liebig

Agricultural fields have been fertilized for thousands of years (Hofstee 1985; Netting 1993: 43), using a range of methods: the application of manure, crop rotation, the inclusion of clover, bringing deeper layers of rich soil to the surface or shipping guano from Chile and Peru to Europe. There was not always a proper scientific understanding of these practices, just as many of the principles underlying ballistics or sailing were not fully understood (see box 5.1). "Proto scientists" such as Thaer and Bousignault derived important insights from these practices, including their theory on humus (the importance of organic material in the soil) and the crucial notion that plants derive many of their building blocks from the air (notably CO_2). Liebig made an additional step by suggesting — and proving — that plant growth critically depends on minerals, especially nitrogen, phosphorus and potassium. He also formulated the theory of the minimum, in which growth factors (such as the presence of different minerals) are represented as unequal staves of a barrel (see figure 5.1). The shortest stave determines the water level in the barrel, i.e., the yields of the crop.

I do not want to understate the importance of Liebig's contribution. On the contrary. The point I want to make here is that both his discovery and the production and application of chemical fertilizer that followed seven decades later, were only possible due to interactions with farming practice. Without the already widespread notion that fertilization mattered, without the many and richly checkered fertilization practices and without the work of many "proto scientists," Liebig's work would have been impossible. And without plant breeders (most of them farmers and later on specialists who built on farmers' skills and practices) who subsequently developed new varieties that could take up higher levels of minerals (notably nitrogen), this discovery would have been useless. When chemical fertilizer became available, there were many alternative trajectories at farmers' disposal — especially approaches that built more directly upon farming practices and within which soil biology played a key role. Rothamsted in the U.K. was one important centre involved in exploring the feasibility of such approaches. It is intriguing that it took another world war to block these alternatives and make chemical fertilization hegemonic.

Roudart [2006] to refer to the introduction and dissemination of tractors) represented another major breakthrough in the development of agricultural productivity. A considerable number of the many mechanical devices associated with this process of motorization have been designed and constructed by farmers themselves. It is telling that the devices developed by farmers frequently embodied design principles that differ significantly from industrial ones. For instance, weeding technologies developed in scientific and/or industrial laboratories aim at reducing labour requirements, whereas weeding technologies designed by farmers are built on the assumption that the available labour force is to be put to best possible use. This difference created strongly contrasting machines and inputs, but also led to highly differentiated substitution curves and emerging product qualities.

The dominant agrarian discourse associates motorization with the notion that "bigger is better," which has led to an ongoing "arms race" in industries producing agricultural technologies. However, the heaviest and most powerful tractor is, in most situations, definitely not the best one. The development of the Italian Ape[8] (a three wheeled vehicle with a light 20 hp motor) did far more for the development of Italian agriculture than the availability of heavy tractors. It not only allowed the farmer to bring the harvest home but could also be used to go with his wife to Holy Mass, the market and the local bar.

Finally there is the case of the high yielding varieties developed in the context of the Green Revolution. In contrast to the small but steady yield increases that are repeated year after year, agricultural engineers often aim to create sudden and considerable leaps that appear as breakthroughs. However, as suggested by Bennett (1982), these jumps might be overtaken after, say, a ten year period, when the yields of "traditional" varieties may exceed those of the "improved" ones. After the leap, the "improved" varieties often remain fixed at the same yield level or even go down slowly. This happened with many of the high yielding varieties that were at the core of the Green Revolution: today, according to many experts, they are "exhausted." Similar episodes can be found in animal production. The "Holsteinization" of European dairy cattle breeds produced a significant jump in milk yields per cow, causing enormous reductions in the size of traditional herds. However, after two decades, those who stubbornly continued breeding with, for example, Friesian cattle achieved equal and sometimes even superior milk yields to those using Holstein cows.

The Green Revolution approach to engineering nearly always embraces a wide array of ingredients, that is, partial changes, which are mutually interdependent. These might include: adjustments to spatial and temporal aspects of farming, major changes in the architecture of plants or animals, the replacement of internal resources and associated farming practices by external inputs and a standardization of fields, practices, norms and parameters. For example, the Holsteinization of herds involved significant changes in temporal rhythms; production that once took place over a long productive lifespan has now become concentrated in a few years. This had a price: longevity has been greatly reduced, and, ironically, it now takes more cows to produce the same amount of milk over, say, a five year period than it did previously.

Science plays an important role in the development of productive forces (Bernstein 2010b), both in general terms and within agriculture. Chayanov was explicit about this point. However, it cannot be maintained that agricultural sciences by definition contribute to developing productive forces or that they are the only force for such development. The picture is far more complicated. A close examination of key episodes from the history of agricultural sciences (such as those described above) shows far more ambiguity than suggested by the "first narrative." Some of these developments came with a price that we are still paying — one that is never taken into account when the hegemony of agrarian sciences over farmers' knowledge systems is asserted.

A voluntary and well supported interplay of farmers' searches for novelties and scientific research can be a powerful driver of agrarian growth and development. History has shown many examples of this; the proposal for a social agronomy as formulated by Chayanov and the current agroecology movement (Altieri et al. 2011) is just one. However, the institutional embedding of agricultural research and theory building increasingly means that they have become constituent parts of "imperial science" (Scott 1998). Science claims to be decisive, but becomes imperial when it reduces agriculture to the sheer application of scientific laws and seeks to standardize, predict, quantify, plan and control agricultural practices. In so doing, it paves the way for farming to be subjected to external prescriptions and control — for food empires to subordinate farming (Vanloqueren and Baret 2009).

A typical feature of imperial science is that it tries to increase agricultural productivity through the construction of new artifacts. These take the form of external resources that add to or replace the resources already available. In contrast, classical agronomy typically looked at the improvement of internal resources — just as is the case in agroecology today. Chemical fertilizers versus the improvement of manure — this exemplifies the contradiction that currently divides agrarian sciences. The conversion of animal excrement into "well-bred" manure (a crucial part of *art de la localité*) is too heterogeneous for imperial science, since the practices will differ from one place to the other. It is also far too capricious, as it depends on many unpredictable and highly variable factors. Manure cannot be controlled from a distance. The same is true of the soil and its biology, intercropping, green manures, the female line in animal breeding, subterranean flows of water and many other aspects of living nature. Manure — like all these other examples — is not a commodity. It is not produced for sale. Hence, it is not of interest to agribusiness.

Imperial science promotes processes of commodification and constructs the instruments that allow for external control. Thus, there is a structural parallelism between the growth and influence of imperial science and those of food empires, which constantly reinforce and reproduce each other. Wherever and whenever imperial science becomes dominant, its contribution to the development of productive forces becomes secondary. Instead its main focus is to contribute to the introduction, extension and consolidation of control (as clearly shown in the case of GMOs). Current agrarian sciences are also biased toward an increased use of fossil energy, just as they are biased toward "optimal conditions" (such as flat fertile lands, large plots, unlimited availability of water, energy and capital and other material inputs), that is, the conditions found on a typical research station. This leads to the development of technologies that do not function so well under suboptimal conditions, which in turn most likely accelerates the marginalization of areas facing such conditions. In addition to this, Stoop (2011: 453) points to

> breeding programmes bypassing a whole set of critical and intricate processes related to the interdependency between

below- and above ground plant parts, that is, between roots and canopy. Likewise, agronomic research has largely bypassed the (micro-) biological and dynamic aspects of the soil and its various interactions with plant roots.

In short, there are very different views about agronomy (see Sumberg and Thompson 2012), and agrarian sciences cannot possibly claim to be free of controversy (Sumberg et al. 2013).

Can Peasants Feed the World?

Again, the answer to this question can be relatively short, since the discussion of Chayanov's work has already identified the major factors. As indicated in chapter 2, peasant agriculture can enter where capital cannot go (in this respect peasant agriculture is "anaerobic" as Raul Paz [2006] phrased it). It goes to the altiplanos of Peru and Bolivia, steep slopes and wet areas elsewhere, the *bolanhas* of western Africa and the *baldios* in the north of Portugal where the costs of cultivation would be far too high to provide even an average return on capital. Such areas are not attractive to capital. Large areas of the world fall into these categories. Much of this land is extensive pasture, especially apt for cattle breeding. Under the aegis of the industrial grain-oilseed-livestock complex (Weis 2007 and 2010), cattle breeding and the production of milk and meat increasingly take place in feedlots where cattle are fed with soy and maize cultivated on fertile, arable land. In a world that increasingly needs this arable land to produce grains to feed the growing human population, this is a ludicrous and unsustainable situation. In short, capitalist agriculture induces counterproductive spatial patterns for the division of labour, whilst it simultaneously degrades the land. On the other hand, in peasant agriculture such distortions are mostly absent.

Second, Chayanov argued that peasant farming is strong in capital formation. The investments per unit of land tend to be higher in peasant agriculture than in capitalist agriculture (this was later shown to be the case in the well-known CIDA studies from the 1960s). To this we can add a third difference: the highly contrasting objectives that orient the different farm types, namely optimization of labour income versus maximization of profit or profitability. A consequence of this is that yields are often higher in peasant agriculture than in capitalist agriculture.

To these "classical" ingredients we can add some additional ones that are becoming evident in the current situation. A fourth factor is that peasant agriculture not only enters where other types of agriculture don't go — it also stays when other forms of agriculture leave (Johnson 2004). This has become very clear in the current period, which is characterized by greater market volatility. Volatility means that market prices suffer huge fluctuations. Low prices may provoke negative cash flows within an enterprise, especially when cost levels are relatively high and cannot readily be changed in the short term. (When a price fall of 40 percent is projected onto figure 5.3, one immediately sees that smallholders in the first column will be badly affected but able to continue farming, albeit with a lower labour income. By contrast, those in the fourth profile will receive a negative return on the capital invested). Therefore capitalist farms will either be closed down or temporarily deactivated, a common phenomenon in large parts of the world. On the other hand, peasant farms are often engaged in economic activities apart from farming (in chapter 6 I will discuss this as multifunctionality). These help them to survive during low price periods. In short, peasant farms are far more resilient than capitalist farm enterprises.

Fifth, peasant farms are far more able to establish the combinations of resources that are most appropriate to local conditions, thanks to the *art de la localité* they have developed (see box 5.2). Having an intimate knowledge of the local ecosystems (Conklin's study of 1957 is still a milestone in this respect), their fields, the available seed material and the individual animals allow peasants to find the locally best fitting solution. The managers of capitalist farm enterprises lack this kind of overview and in-depth knowledge. By necessity they operate scientific schemes that are, by nature, standardized and view local details as systemic coarseness.[9] This can result in far higher levels of emissions and other types of losses and in less than optimal resource use.

A sixth ingredient, which builds on the previous one, is novelty production, which allows peasant agriculture to develop the resources on which it is based. This is particularly relevant given the variety that exists within the peasant farm and its fields (see, for example, Brush et al. 1981).

The fifth and sixth ingredients flow together into a seventh one: peasant agriculture is, mostly, more sustainable than capitalist agriculture. It is more rooted in the local ecosystems (see the discussion on coproduction in chapter 3) and thus more resistant to events such as droughts; it is less dependent on fossil fuels (Ventura 1995; Netting 1993: 123–45); its animals are mostly longer lived; there is intercropping that can give additional synergies (with the residuals often being reused); it helps to avoid climate change (Altieri and Koohafkan 2008); and finally seeks to minimize water losses (Dries 2002). Consequently, peasant agriculture is not only well-equipped to face the huge challenge of feeding the world — it is also able to contribute considerably to addressing these "new scarcities" and climate change. It also generates productive and socially and individually meaningful employment, far more than capitalist farm enterprises (or the cities for that matter)[10] can ever provide. Finally, peasant agriculture also helps to create dignified work and livelihoods

For over three decades I have been working, on and off, with a team of Italian and Dutch colleagues documenting the performance of a group of peasant-like farms and a second group of entrepreneurial farms (whose style of operation is close to that of capitalist enterprises) in Parma, Italy. Both groups specialize in dairy farming and operate under similar conditions. In order to construct figure 5.5, the specific features of each group (differences in size, labour input, investments, technical efficiency, cattle density, longevity, yields, overall level of production per hectare) have been translated to an imaginary block of 1,000 hectares in order to allow a comparison between the two contrasting ways of farming. Thus, the figure shows the overall production that would be realized under the two contrasting approaches.

The differences are striking. In 1971 peasant farming would yield 15 percent more than the entrepreneurial mode. This difference has steadily increased over time. In 1999 peasant production yielded 56 percent more, and by 2009, the figure was nearly twice as high (partly because many entrepreneurial farms were deactivated).[11]

Figure 5.5 Comparison of Entrepreneurial and Peasant Farming in Parma, Italy

	Entrepreneurial farming	Peasant farming
GVP in 1971	735 million lire	844 million (+15%)
GVP in 1979	2,845	3,872 (+36%)
GVP in 1999	8,235	12,815 (+56%)
GVP in 2009	5.4 million euro	10.7 million euro (+98%)

These differences can be attributed to a wide range of details. Often they are small details (such as longevity and productivity of cows, grassland productivity, etc.). While these mostly go unnoticed, together they create a significant difference. The entrepreneurial model farms are mostly larger than the peasant units. They look more impressive and are more mechanized — all signs that are mostly translated into "more powerful" and "more competitive." However, appearances are deceptive. Although a single entrepreneurial enterprise produces more than a single peasant unit, a thousand hectares of land used by peasant farms produces far more than the same thousand hectares used by entrepreneurial or capitalist farms.

Can peasant farms feed the world? Yes they can. And they could do it even better if we could curtail the amount of added value that is now siphoned off by food empires (Polayni 1957; Friedmann 2004). If these empires appropriated less (or none) of the value produced in peasant units, and if peasants could have access to more of the best arable land, the labour incomes in the peasant farms would increase, allowing for more capital formation and further development and growth. The answer could also be more affirmative if the inbuilt biases in agrarian sciences were corrected so that they related in an appropriate way to the peasantries of the world, as exemplified by the social agronomy proposed by Chayanov.

Notes

1. The title in Russian refers to the "labour farm." Previous drafts seem to suggest that this change, from "peasant farm" to "labour farm," was a last-minute one. The change was probably due to the polemics and tensions that later on resulted in Chayanov's tragic deportation and death. Sadly, history seems to have no memory. Many decades later, the Brazilian military government (from the early 1970s) officially banned the word *peasant*. It reminded people too much of the Legas Camponesas (Peasant Leagues) that had been brutally repressed by the same military.
2. This observation was one of the important cornerstones of the "social agronomy" developed in the 1930s and 1940s in northwestern Europe and some of its colonies. It facilitated the understanding of how social and agronomic aspects flow together in one single process of coproduction and co-evolution (see Timmer [1949] and Vries [1931] who both strongly built upon Chayanov). Thus an integration of social sciences and agronomy in one single "social agronomy" becomes theoretically feasible. Currently, agro-ecology might be considered as the continuation and further improvement of this trajectory.
3. The image contained in figure 5.1 is one used since Liebig. It is very helpful for didactic purposes. However it does not account for multiple interactions and synergies between specific growth factors.
4. The impact of novelties has been expressed through the concept of X efficiency (Yotopoulos 1974). X efficiency describes a superior economic performance, in which the economic results exceed those that can be explained by the available factors of production and technology. X efficiency is the "unknown part" (hence the X). Novelties are a decisive ingredient in creating X efficiency. They can make the economy perform better, drive the "frontier function" in an upward direction (Timmer 1970) and are decisive in "disembodied technological change" (Salter 1966).
5. This is an important contrast with the photo-insensitive short-straw cultivars that were at the heart of the Green Revolution. "Modern" rice cultivation, as defined in and by the Green Revolution, involved a shift away from solar energy and human labour toward a strongly increased use of fossil energy in the form of chemical fertilizer. SRI builds again on soil biology, solar energy and local knowledge.

6. The crucial requirements here are that the peasant has enough resources to buy twenty euros worth of inputs, that weather conditions permit the crop to develop well and that his irrigation water is not taken by more powerful others.
7. These three examples correspond to the main mechanisms of what Mazoyer and Roudart (2006: 375) refer to as the "second agricultural revolution of modern times." These are: motorization and mechanization; synthetic fertilizers; and seed selection (see Mazoyer and Roudart 2006: 375ff.).
8. Ape literally means bee (just as Vespa, the symbol of urban mobility, means wasp).
9. A typical example here is standardized fertilizer doses (e.g., 400 kg N per hectare) versus variations within a field related to different degrees of soil fertility.
10. I refer here to the inability of cities to absorb the rural population that becomes superfluous when farming is reorganized on a capitalist basis.
11. Differences such as these are often camouflaged. A particular feature of dairy farming in Parma province is that it is linked to the production of Parmesan cheese. Hence, no silage may be used. This means in practice that all, or nearly all, the roughage (grass and hay) is produced within the farm itself. The numbers of cattle per hectare cannot fluctuate too much. This is a basic difference with dairy farming elsewhere. In other regions the intensity is often a function of the feed and fodder bought from elsewhere. The Netherlands, for instance, has an agricultural area of some 2 million hectares. But Dutch agriculture makes use of some 16 million hectares of land outside its borders. This land is mostly used to produce feed and fodder (notably soybeans and maize) that is imported into the Netherlands and used for feeding livestock.

6

Repeasantization

In 1978 peasants from Xiaogang, a small village in Anhui province, China, agreed that it was impossible to continue working according to the commune system. The system degraded them and hunger was the only destiny they could see. They concluded that they would be better off begging rather than continuing to farm in this way (Gulati and Fan 2007). This led them to secretly decide to contract the production team's land to the individual peasant families and let them work these plots according to their own capabilities and needs (i.e., according to their specific labour-consumer balance). This did not imply rejecting the principle that, as part of the agricultural sector, they should contribute to the nation and its development. One of their banners clearly stated that they were willing to contribute to the state and the collective. However, "all that is left is ours" (Wu 1998). The eighteen peasants involved signed a secret document in which they promised to take care of each other's children if some of them were killed or imprisoned. The contract was a typical peasant document in that it specified that this commitment only applied until the children were eighteen years old. Peasants never commit to unnecessary spending.

This is how *da baogan* (an expression that is extremely vague but can be loosely translated as "big contracting to you"[1]) began. After interventions and help, first from regional party authorities and later from Deng Xiao Ping, it developed into what became known as the Household Responsibility System (HRS). HRS is understood by those who helped to introduce and generalize it as an "institutional innovation" and a "transformation" that restored "micro-level agency" (Du 2006: 2 and 11). The HRS made the Chinese peasants visible once again on the national scene. The state-controlled management of farming by people's communes was replaced by the individual decisions of peasant households.

This form of repeasantization induced an enormous increase in agricultural production:

> Agricultural output increased by 42.2% in the period from 1978 to 1984 (calculated using fixed prices); 46.9% of this growth could be attributed to the systems reforms, 32.2% to increased fertilizer use and the rest to other factors. This increase in agricultural production led to previous problems of food scarcity being resolved in a short time and the number of impoverished people was reduced from 250 million (30.7% of the population) in 1978 to 21.5 million (2.3%) in 1990. (Ye et al. 2010: 263–4; see also Deng 2009 and Li et al. 2012)

Netting (1993: 252) adds that "per capita incomes went up even faster than production, by 102%, and indices of living standards like average floor space climbed almost a third to 13.41 square metres."

In the spring of 2012, I had the opportunity to talk at length with two peasants from the initial group of eighteen: Yan Hongchang and Yan Jinchang, both of whom are still farming. The issue of yields plays an important role in their explanations. They told me,

> At that time we cultivated 300 *mu*, but production only reached some 20,000 *jin*.[2] One part of this was to be reused as seed, another part was for public purposes and the final part was for ourselves. But it was not enough ... If you plant 20 *jin* of seed and you only harvest 60, then something is very wrong. And we knew it could be different. After the first land reform [1951], our parents produced far more on the same land and in 1962, during the emergency, we noted again that far more could be produced here. But in the commune system [from 1959 onwards] total production declined. There was no motivation among farmers to work hard, we got depressed, we could not feed our families anymore, life lost its meaning. Seeing bad yields makes you feel useless and guilty.

> When we started working as peasants again, we were able to realize high yields. We even gave the state far more than our quota. That was because we wanted to give a good impression to the state in order to get support ... Having the right to make

decisions was very important for us. Individual motivation is a driving force, when you have your own land you care better for the plants ... All this is evident; when doing farm work the purpose is to get good results. (Hongchang and Jinchang, pers. comm.)

There are also glimpses of balancing in the memories of these two veterans. They explained to me that in agriculture "one gives and one gets" (ibid.). And that after suffering "pain" there will be "gain" (there could hardly be a more precise description of the balance between drudgery and satisfaction). "Only when you make an effort will the fields yield good harvests and only then you'll get the benefits" (ibid.). On the other hand, "it isn't fair if you don't get the results of your hard work" (ibid.).

Another important balance that has been actively reconstructed since the 1980s is that of town-countryside relations, principally mediated by migration patterns (as described in chapter 4 of this book). The circular character of labour migration in China (people leave the village but come back later in order to continue farming) strengthens rather than weakens peasant farming (Ploeg and Ye 2010).

Processes and Expressions of Repeasantization

The transition of Chinese agriculture from collective to peasant agriculture is just one example, albeit an important one, of current trends towards repeasantization. Repeasantization can take many different paths (Enriquez 2003; Rosset and Martínez-Torres 2012). Another example is the Movimento dos Trabalhadores Sem Terra (Landless Workers' Movement or MST) in Brazil that resulted in the creation of more than 400,000 new peasant units (Veltmeyer 1997). Following Victor Toledo (2011) the agroecological movement can also be described as repeasantization. The same applies in Eastern Europe where new strata of peasants emerged from the transitions of the 1990s and are struggling to build up new agricultures (Spoor 2012). Equally significant is the rise of Via Campesina (the Peasants' Way), a new and proud movement that heralds the possibility and promise of peasant agriculture regaining a central role in global agriculture (Desmarais 2002; Borras 2004). Via Campesina's role in the main socio-political struggles and its persistence in addressing UN organizations such as the FAO is an expression par excellence of the trend toward repeasantization.

I cannot possibly discuss all these trends and expressions in this small book. Nor is there any need to do so. Much of the information is easily accessible. I will, though, make one exception. And that is to talk about processes of repeasantization in western Europe. I will do so because many people are still uncomfortable with the interpretation of the changes currently occurring in western European agriculture as representing a process of repeasantization.

Repeasantization in Western Europe: Resetting the Balances

Within the European Union a minority of farmers (some 15 to 20 percent) follow the "entrepreneurial road," centred on accelerated scale increases, technologically driven intensification and a tightening of dependency relations with the food industries, banks and retail chains. From one perspective this is logical. The "agricultural entrepreneurs" are already locked into this system, through high levels of indebtedness and input use. They are, as it were, entrapped. For them, there is only one way ahead. On the other hand they pay a high price in following this route. Remuneration for their long, monotonous and, sometimes, dangerous work is low. In times of crisis it is negative. Although the balance of labour and consumption is not completely broken, it can be extremely difficult to create a satisfactory equilibrium. Particular problems can arise when a son and/or daughter (and their families) want a share in the enterprise; they have to engage in risky financial operations, involving huge loans. Sometimes the balance is achieved in another way: by contracting badly paid, "black" workers (from Poland, India, the Maghreb or sub-Saharan Africa, for example). Similar uncertainties apply to the balance between drudgery and utility. Here a specific equilibrium is created by redefining the very notion of utility. Their utility is located somewhere in the future: they believe that, as large farmers, they will be among the few that survive, that accelerated growth is the surest way to ensure their competitiveness in the future.

In contrast, the majority of farmers follow a different route. They reassess the main balances in completely different ways and in doing so, they are making a large part of European agriculture more peasant-like. They face the squeeze on agriculture, which is particularly acute at the moment, by reassessing the balance between internal and external resources (see chapter 3). They reduce their dependency on external resources (including credit) and seek to optimize the use of internally available resources. This reduces their financial and transaction costs, while increasing their labour income — at a given level of total production. We are not talking about marginal improvements — or a "few kernels of grain" — here. Long-term comparative research in the State Research Centre for Dairy Farming in the Netherlands shows that a low cost farm producing 400,000 litres of milk can earn the same income as a high-tech farm that produces 800,000 litres (Kamp et al. 2003; Evers et al. 2006). The labour input on these two farms is the same. This means that, at a given level of production, labour income can be doubled by shifting from a high-tech style of farming toward a low cost one. Reassessing the balance of internal and external resources might take quite some time, and it might also affect the other balances. Coproduction, for instance, might be grounded more on living nature, which makes it easier to integrate caring for the landscape, nature and biodiversity into farming practices. This, then, can improve the balance between the farming family and their neighbourhood. This latter balance is one that entrepreneurial farmers find increasingly problematic to maintain.

A second major ingredient of the peasant trajectory is the development of multifunctionality; new products and new services are being produced and increasingly marketed through newly constructed nested markets. Here again "the family farm uses, within its power, all the opportunities of its natural and historical position and of the market situation in which it exists" (Chayanov 1966: 120). These activities are undertaken to increase labour income. There are a vast range of such activities and opportunities in Europe: agro-tourism, high quality products, regional specialties, organic production, on-farm food processing, direct selling (many different methods have been developed), energy production, water storage, care facilities, housing horses, management of landscape and nature and many other forms of diversification. At the end of the 1990s such new activities within the EU generated an additional income to labour of more than 8 billion euros (twice the total annual agrarian income of the Netherlands). This has allowed millions of small and medium family farms to survive (data from Ploeg, Long and Banks 2002). Novelty production adds considerable momentum to these new activities. There is a multitude of European farmers involved in these activities: a newly emerging peasantry. It is "an ensemble of singularities ..., it is productive ... always in movement" (Negri 2008) and it contains creative power. The shortcomings of food empires and state apparatuses create many interstices (or gaps) that this multitude uses as points of departure to create new practices that perform better. In the long run, these might result in important changes in the politico-economic contours of agriculture. In this respect, José Bové, the French peasant leader, observed that "if you add together the various initiatives ... you begin to get a strong feel of a new farmers' movement which, I believe, will eventually marginalize industrial agriculture (Bové and Dufour 2001: 42).

Like the shifting balance in the use of internal and external resources, this newly constructed multifunctionality involves more than a few kernels of grain. All the available studies show that these newly wrought productive activities contribute considerably to rural incomes, at both the farm and regional levels (Heijman et al. 2002). They make a large contribution to sustaining farms that might otherwise disappear or be coerced to follow the entrepreneurial road. Of special interest here is the development of new markets that are nested in new arrangements between producers and consumers (Ploeg, Ye and Schneider 2012). These new markets can be considered to be commons. The construction of such commons is not restricted to Europe. In China and, especially, in Brazil highly innovative forms of newly constructed markets have emerged and are growing (see Ye et al. 2010; Schneider, Shiki and Belik 2010; Perez 2012).

These new forms of repeasantization (also discussed in Brookfield and Parsons 2007) critically involve a rebalancing of the equilibrium between drudgery and utility. Those building new, multifunctional farms, grounded on a relatively autonomous resource base, are coming to redefine drudgery. Such farmers mention working outside, highly diversified tasks, independence and working with living nature as being among the more attractive aspects of their work. They experience far less drudgery than those following the entrepreneurial path, where labour can be monotonous, risky and dull. Utility is also experienced in a different way. Alongside good earnings, there is the joy of meeting far more people (entrepreneurial farmers typically experience high levels of solitude) and the pride of "farming differently" (Oostindie et al. 2011). These are now becoming important ingredients of the utility experienced by Europe's new peasantry. This provides a further impetus for shifts such as those shown in figure 2.1 and further strengthens the emergence of a new peasant-like agriculture.

Thus, in the heart of one of the most modernized agricultural systems in the world, we can still see the mechanics Chayanov described almost a century ago. Different balances matter. However, in the modern world, such considerations are no longer limited to the peasant family — society at large is increasingly involved in assessing these balances, which means there are links between farming and different social arenas. This helps the emergence of different ways of setting the balances, allowing for different routes, such as the peasant and the entrepreneurial pathways and others. In short, balancing is still central to farming. But it can be for the better or for the worse. One set of balances helps to shape entrepreneurial trajectories that are increasingly at odds with contemporary social expectations. Another set of balances can help shape new routes for repeasantization that have a completely different impact. World agriculture is, indeed, at a crossroads, and insights into these strategic balances are needed more than ever before in order to understand the dilemmas and devise the most appropriate solutions.

Notes

1. This vagueness is an intriguing feature of many experiments and politico-economic changes in China. It helps to avoid premature or unnecessary conflicts.
2. A *jin* is half a kilogram. In this context it means that 10,000 kg of wheat and rice were produced. A *mu* is 1/15 hectare.

Glossary

agrarian question: The issues that arise when there is a serious disturbance of the relations between, on the one hand, the way in which farming is organized and, on the other, ecology, society and/or the interests and prospects of those directly engaged in agricultural production

capital: Value that is used to obtain surplus value; capital requires wage labour.

capitalism: A distinctive and globally established socio-economic system, based on the class relation between labour and capital

commoditization: The process that results in the elements of production and reproduction being produced for and obtained from market exchange, making them subject to its logic

commodity: A product or service that is produced for and/or obtained through market exchange

commons: Jointly owned assets (including nonmaterial ones) that can be used to create more value. Commons differ from capital in that they do not have to render surplus value, nor do they function as commodities.

coproduction: The interactions between man and living nature, which lead to the mutual transformation of both. Coproduction can involve both ecological and market exchanges and is a very important aspect of agriculture.

corporate agriculture: A form or mode of farming fully based on wage labour; usually large-scale; its main internal driver is to obtain the highest possible return on capital.

depeasantization: The loss or disappearance of the peasantry. This occurs through a variety of processes that impede peasant farmers' access to the means to reproduce their peasant way of farming.

desjatína: A Russian unit of area measurement; 2.7 acres or 1.1 hectares.

downstream markets: The markets where agricultural commodities are sold when they leave the farm

drudgery: The effort required to produce a product or service; it is assumed that the extra drudgery needed to produce an extra unit increases in line with total production.

ecological exchange: The interaction between a unit of production (e.g., a farm) and the surrounding ecosystem; this form of exchange is noncommodity based.

entrepreneurial agriculture: A form or mode of agriculture in which market exchange is much stronger than ecological exchange; its resource base is highly dependent on external actors (e.g., banks). It often expands by taking over the resources of other farmers.

exploitation: The appropriation of the surplus product of classes of producers by (dominant) classes of nonproducers

extensionists: Professionals trained in the communication of innovations to farmers

external resources: Those resources acquired from upstream markets that enter the process of production as commodities and thus bring the logic of the markets into the heart of the process of production

farmers: A generic term that refers to those actors actively involved in the agricultural labour process; these might be peasants, entrepreneurial farmers, agricultural workers, etc.

farming: A generic term that embraces peasant agriculture, entrepreneurial agriculture and corporate agriculture

food empire: An extended network that exerts oligopolistic control over the production, processing, distribution and consumption of food and that simultaneously appropriates a large proportion of the value produced from these activities

food regime: The international system of relations, rules and practices that structures the production, processing, distribution and consumption of food; the current food regime is often characterized as the corporate or imperial food regime.

gender relations: Relations between men and women; the gender division of property, labour and income is typically unequal.

globalization: Widely considered to be the current phase of world capitalism, especially from the 1980s onwards. Its effects are much debated, but it is characterized by largely unregulated internal capital markets, the dominance of financial capital and the political project of neoliberalism.

instruments of labour: Tools used to facilitate and/or improve the process of labour. They might be simple or sophisticated; in peasant studies (sophisticated) instruments are often erroneously equated with capital.

intensification: A process that aims at and results in ongoing yield increases

internal resources: Those resources produced and reproduced within the unit of production

kolkhoz: large, state managed agricultural enterprise that characterized Russian agriculture during the communist epoch

labour income: The return acquired from the products and services sold, minus the monetary costs required to produce these products and services

labour process: The organization and activities of labour in production processes

labour productivity: The amount of a good or service someone can produce with a given expenditure of effort, typically measured in terms of time spent working or labour time

market exchange: The interaction between a unit of production (e.g., a farm) and the upstream and downstream markets. This type of exchange involves commodities.

neoliberalism: A political and ideological program to "roll back the state" in the interests of the market and its major capitalist actors

noncommodity: A product or service not obtained through the market but created within the productive unit itself that is used within the process of production and/or a product or service obtained through socially regulated exchange

objects of labour: Those ingredients of the labour process that are converted into new products that represent an increased value (e.g., fertile fields, dairy cows and fruit trees)

peasant agriculture: Forms or modes of farming in which coproduction based on a self-controlled resource base is central and within which wage labour is (almost) absent. Enlarging the value added per object of labour is an important internal driver for its development.

peasantries: Conglomerations of peasants sharing common experiences and identities, making use of internal mechanisms to exchange ideas and resources and to attribute authority to leaders. They share joint notions about how farming should be organized and developed and may also share and/or jointly develop commons.

peasants: Social actors engaged in peasant agriculture

petty commodity producer: An analytical term mostly used to refer to those who use forms or modes of production oriented to the market but based on noncommodity resources and relations. Peasant agriculture is a form of petty commodity production.

primitive accumulation: For Marx the historical processes through which the key classes of capitalism are established. It also describes processes that rely on coercive extra-economic mechanisms to squeeze out as much wealth as possible from particular classes. More specifically it has been used to describe increases in the exploitation of the peasantry that were used to accelerate industrialization.

production: The process through which labour is applied in changing nature to satisfy the conditions of human life.

productivity: How much can be produced with a given amount of resources (land, labour, water, etc.)

pud: A Russian weight, 16.4 kilograms

repeasantization: The process through which agriculture is restructured as peasant agriculture. It may also refer to a quantitative increase of the numbers of peasants.

reproduction: Securing the conditions of life and of future production from what is produced or earned now

resources: The social and material elements needed to sustain the process of production (land, labour, knowledge, animals, plants, networks, etc.). The required resources might be produced and reproduced within the unit of production, obtained through socially regulated exchange and/or purchased from upstream markets.

self-controlled resource base: Allows for relative autonomy as it is largely, though not completely, based on the production and reproduction of resources within the unit of production

social relations of production: All those social relations, institutions and practices that shape the activities of production and reproduction and that simultaneously regulate the distribution of the wealth produced

squeeze on agriculture: Unfavourable exchange relations (stagnating or decreasing off-farm prices and increasing costs) that drain wealth from agriculture and that increasingly threaten the reproduction of both the farm and the farming family

upstream markets: Those markets that can provide the resources needed for farming, e.g., land, labour, instruments of labour, all kind of material inputs, credit, etc.

utility: The sum of values (of both a commodity and a noncommodity nature) that results from the process of production

yield: Measure of the productivity per labour object; usually the amount of a crop harvested from a given area of land and/or the amount of products produced per animal.

References

* Signifies recommended further reading of an introductory kind
** Signifies recommended further reading of a more advanced kind

Abramovay, Ricardo. 1998. "O admirável mundo novo de Alexander Chayanov." *Estudos Avançados* 12, 32.

Adey, Samantha. 2007. *A Journey Without Maps: Towards Sustainable Agriculture in South Africa*. Wageningen, The Netherlands: Wageningen University.

Agarwal, Bina. 1997. "Bargaining and Gender Relations Within and Beyond the Household." *Feminist Economics* 3, 1.

Altieri, Miguel A. 1990. *Agroecology and Small Farm Development*. Ann Arbor, MI: CRC Press.

Altieri, Miguel A., Fernando R. Funes-Monzote and Paulo Petersen. 2011. *Agroecologically Efficient Agricultural Systems for Smallholder Farmers: Contributions to Food Sovereignty*. Paris/Berlin: INRA and Springer-Verlag.

Altieri, Miguel A., and Parviz Koohafkan. 2008. *Enduring Farms: Climate Change, Smallholders and Traditional Farming Communities*. Penang, Malaysia: TWN, Third World Network.

Arkush, D. 1984. "'If Man Works Hard the Land Will Not Be Lazy': Entrepreneurial Values in North Chinese Peasant Proverbs." *Modern China* 10, 4.

Auhagen, O. 1923. "Vorwort." In A. Chayanov (Tschajanow) *Die Lehre von der bäuerlichen Wirtschaft, Versuch einer Theorie der Familienwirtschaft im Landbau*. Berlin: Verlagsbuchhandlung Paul Parey.

Bagnasco, A. 1988. *La Costruzione Sociale del Mercato, studi sullo sviluppo di piccole imprese in Italia*. Bologna, Italy: Il Mulino.

Ballarini, G. 1983. *L'animale tecnologico*. Parma, Italy: Calderini.

Barrett, C.B., T. Reardon and P. Webb. 2001. "Nonfarm Income Diversification and Household Livelihood Strategies in Rural Africa: Concepts, Dynamics, and Policy Implications." *Food Policy* 26.

Bennett, John W. 1982. *Of Time and the Enterprise, North American Family Farm Management in a Context of Resource Marginality*. Minneapolis: University of Minnesota Press.

Benvenuti, B. 1982. "De technologisch administratieve taakomgeving (TATE) van landbouwbe-drijven." *Marquetalia* 5.

Benvenuti, B., S. Antonello, C. de Roest, E. Sauda and J.D. van der Ploeg. 1988. *Produttore agricolo e potere; modernizzazione delle relazioni sociali ed economiche e fattori determinanti dell'imprenditorialita agricola*. Rome: CNR/IPRA.

Benvenuti, B., E. Bussi and M. Satta. 1983. *L 'imprenditorialitá agricola: a la ricerca di un fantasma*. Bologna, Italy: AIPA.

** Bernstein, Henry. 2010a. *Class Dynamics of Agrarian Change*. Halifax: Fernwood Publishing.

____. 2010b. "Introduction: Some Questions Concerning the Productive Forces." *Journal of Peasant Studies* 10, 3.

____. 2009. "V.I. Lenin and A.V. Chayanov: Looking Back, Looking Forward." *Journal of Peasant Studies* 36, 1.

Bieleman, J. 1992. *Geschiedenis van de landbouw in Nederland, 1500–1950*. Meppel, The Netherlands: Boom.

Boelens, R. 2008. *The Rules of the Game and the Game of the Rules: Normalization and Resistance in Andean Water Control*. Wageningen, The Netherlands: Wageningen University.

Bonnano, A., L. Busch, W. Friedland, L. Gouveia and E. Mingione. 1994. *From Columbus to Conagra: The Globalization of Agriculture and Food*. Lawrence: University Press of Kansas.

Borras, S.M. 2004. *La Via Campesina: An Evolving Transnational Social Movement*. Amsterdam: Transnational Institute.

Borras, S.M., Marc Edelman and Cris Kay. 2008. "Transnational Agrarian Movements: Origins and Politics, Campaigns and Impact." In S.M. Borras et al. (eds.), *Transnational Agrarian Movements Confronting Globalization*, special issue, *Journal of Agrarian Change* 8, 1/2.

Boserup, Ester. 1970. *Evolution agraire et pression demographique*. Paris: Flammarion.

* Bové, J., and F. Dufour. 2001. *The World Is Not for Sale*. London: Verso.

** Bray, Francesca. 1986. *The Rice Economies: Technology and Development in Asian Societies*. Oxford: Blackwell.

Brookfield, Harold, and Helen Parsons. 2007. *Family Farms: Survival and Prospect, a World-Wide Analysis*. Oxford: Routledge.

Brox, O. 2006. *The Political Economy of Rural Development: Modernisation Without Centralisation?* Delft, The Netherlands: Eburon.

Brush, S.B., J.C. Heath and Z. Huaman. 1981. "Dynamics of Andean Potato Agriculture." *Economic Botany* 35, 1.

Bryden, J.M. 2003. "Rural Development Situation and Challenges in EU-25." Keynote speech to the EU Rural Development Conference, Salzburg, Austria.

Cassel, Guilherme. 2007. "A atualidade da Reforma Agraria." *Jornal Folha de Sao Paulo* March 4

Chambers, J.D., and G.E. Mingay. 1966. *The Agricultural Revolution 1750–1880*. London: B.T. Batsford Ltd.

Chayanov, Alexander. 1991 [1927]. *The Theory of Peasant Co-operatives*. Columbus: Ohio State University Press.

** ____. 1988 [1917]. *L'economia di lavoro, scritti scelti*, a cura di Fiorenzo Sperotto. Milan: Franco Angeli/INSOR.

____. 1976 [1920]. "The Journey of My Brother Alexis to the Land of Peasant Utopia." *Journal of Peasant Studies* 4.

** ____. 1966 [1925]. *The Theory of Peasant Economy*. (D. Thorner et al., editors.) Manchester: Manchester University Press.

____. 1966b. "On the Theory of Non-Capitalist Economic Systems." In Chayanov, *Theory of Peasant Economy*.

____. 1924. *Die Sozial Agronomie, ihre Grundgedanken und ihre Arbeitsmetoden*. Berlin: Verlagsbuchhandlung Paul Parey.

____. 1923. *Die Lehre von der bäuerlichen Wirtschaft, Versuch einer Theorie der Familienwirtschaft im Landbau*. Berlin: Verlagsbuchhandlung Paul Parey.

Columella, Luciano G.M. 1977. *L' arte dell'agricoltura e libro sugli alberi*. Torino, Italy: Einaudi editore.

Conklin, H.C. 1957. *Hanunóo Agriculture, A Report on an Integral System of Shifting Cultivation in the Philippines*. Rome: FAO.

Danilov, Viktor. 1991. "Introduction: Alexander Chayanov as a Theoretician of the Co-operative Movement." In Alexander Chayanov, *The Theory of Peasant Co-operatives*. Columbus: Ohio State University Press.

Dannequin, Fabrice, and Arnaud Diemer. 2000. "L'economie de l'agriculture familiale de Chayanov a Georgescu-Roegen." Paper presented at Colloque SFER, Paris, November 2000.

Davis, M. 2006. *Planet of Slums*. London: Verso.

Deléage, Estelle. 2012. "Les paysans dans la modernité." *La Découverte/Revue Française de Socio-Economie* 1, 9.

Deng, Zhenglai. 2009. "Academic Inquiries into the 'Chinese Success Story.'" In Zhenglai Deng (ed.), *China's Economy, Rural Reform and Agricultural Development*. Singapore: World Scientific Publishing Co.

Desmarais, A. 2002. "Peasants Speak — The *Via Campesina*: Consolidating an International Peasant and Farm Movement." *Journal of Peasant Studies* 29, 2.

Domínguez García, M.D. 2007. *The Way You Do It Matters: A Case Study on Farming Economically in Galician Agroecosystems in the Context of a Cooperative*. Wageningen, The Netherlands: Wageningen University.

Dries, A. van der. 2002. *The Art of Irrigation: The Development, Stagnation and Redesign of Farmer-Managed Irrigation Systems in Northern Portugal*.

Wageningen, The Netherlands: Circle for Rural European Studies, Wageningen University.

Du Runsheng. 2006. *The Course of China's Rural Reform*. Washington, DC: International Food Policy Research Institute.

Durrenberger, E. Paul. 1984. *Chayanov, Peasants, and Economic Anthropology*. Orlando: Harcourt Brace.

Edelman, M. 2005. "Bringing the Moral Economy Back in … to the Study of 21st Century Transnational Peasant Movements." *American Anthropologist* 107, 3.

Engel, P.H.G. 1997. *The Social Organization of Innovation: A Focus on Stakeholder Interaction*. Wageningen, The Netherlands: Wageningen University.

Enriquez, L.J. 2003. "Economic Reform and Repeasantization in Post-1990 Cuba." *Latin American Research Review* 38, 1.

Evers, A.G., M.H.A. de Haan, K. Blanken, J.G.A. Hemmer, G. Hollander, G. Holshof and W. Ouweltjes. 2006. "Results Low Cost Farm, 2006, Rapport nr. 53." Wageningen, The Netherlands: Animal Science Group, Wageningen University.

Fei Xiao Tung . 1939. *Peasant Life in China: A Field Study of Country Life in the Yangtze Valley*. London: George Routledge and Sons.

Friedmann, H. 2004. "Feeding the Empire: The Pathologies of Globalized Agriculture." In R. Miliband (ed.), *The Socialist Register*. London: Merlin Press.

*___. 1993. "The Political Economy of Food: A Global Crisis." *New Left Review* 1.

**___. 1980. "Household Production and the National Economy: Concepts for the Analysis of Agrarian Formations." *Journal of Peasant Studies* 7.

Galeano, Eduardo. 1971. *Open Veins of Latin America: Five Centuries of the Pillage of a Continent*. New York: Monthly Review Press.

Garstenauer R., Sophie Kickinger and Ernst Langthaler. 2010. "The Agrosystemic Space of Farming: Analysis of Farm Records in Two Lower Austrian Regions, 1945–1980s." Paper to the Institute of Rural History workshop, Historicising Farming Styles, in Melk, Austria, October 22–23, 2010.

Geertz, C. 1963. *Agricultural Involution*. Berkeley, CA: University of California Press.

Georgescu-Roegen, N. 1982. *Energia e Miti Economici*. Torino, Italy: Editore Boringhieri.

Gerritsen, P.R.W. 2002. *Diversity at Stake: A Farmer's Perspective on Biodiversity and Conservation in Western Mexico*. Wageningen, The Netherlands:

Circle for Rural European Studies, Wageningen University.

Gulati, Ashok, and Shenggen Fan. 2007. *The Dragon and the Elephant: Agricultural and Rural Reforms in China and India.* Baltimore: IFPRI/ Johns Hopkins University Press.

Halamska, M. 2004. "A Different End of the Peasants." *Polish Sociological Review* 3, 147.

Hardt, Michael, and Antonio Negri. 2004. *Multitude: War and Democracy in the Age of Empire.* New York: Penguin Press.

Harvey, David. 2010. *The Enigma of Capital and the Crises of Capitalism.* London: Profile Books.

Hayami, Yujiro. 1978. *Anatomy of a Peasant Economy: A Rice Village in the Philippines.* Los Baños, Philippines: International Rice Research Institute.

Hayami, Y., and V. Ruttan. 1985. *Agricultural Development: An International Perspective.* Baltimore: Johns Hopkins.

Hebinck, P. 1990. *The Agrarian Structure in Kenya: State, Farmers and Commodity Relations.* Saarbrucken: Verlag Breitenbach.

Heijman, Wim, M.H. Hubregtse and J.A.C. van Ophem. 2002. "Regional Economic Impact of Non-Standard Activities on Farms: Method and Application to the Province of Zeeland in the Netherlands." In J.D. van der Ploeg, A. Long and J. Banks (eds.), *Living Countryside: Rural Development Processes in Europe — The State of the Art.* Doetinchem, The Netherlands: Elsevier.

Hofstee, E.W. 1985. *Groningen van Grasland naar Bouwland, 1750–1930.* Wageningen, The Netherlands: Pudoc.

Holloway, John. 2010. *Crack Capitalism.* London: Pluto Press.

____. 2002. *Change the World Without Taking Power.* London: Pluto Press.

Huang, Philip C.C. 1990. *The Peasant Family and Rural Development in the Yangzi Delta 1350–1988.* Stanford, CA: Stanford University Press.

IAASTD (International Assessment of Agricultural Knowledge, Science and Technology for Development). 2009. *Agriculture at a Crossroads: Global Report.* Washington, DC: Island Press.

IFAD (International Fund for Agricultural Development). 2010. *Rural Poverty Report 2011: New Realities, New Challenges, New Opportunities for Tomorrow's Generation.* Rome: IFAD.

Jackson, Tim. 2009. *Prosperity Without Growth? The Transition to a Sustainable Economy.* London: Sustainable Development Commission.

Janvry, A. de. 2000. "La logica delle aziende contadine e le strategie di sostegno allo sviluppo rurale." *La Questione Agraria* 4.

* Johnson, H. 2004. "Subsistence and Control: The Persistence of the

Peasantry in the Developing World." *Undercurrent* 4, 1.
Kamp, A. van der, A.G. Evers and B.J.H. Hutschemaekers. 2003. *Three Years High-Tech Farm, Praktijkrapport Rundvee*, nr. 26. Wageningen, The Netherlands: Animal Science Group, Wageningen University.
Kautsky, Karl. 1974 [1899]. *La cuestión agraria*. Buenos Aires: Siglo Veintiuno, Argentina Editores.
Kay, Cristóbal. 2009. "Development Strategies and Rural Development: Exploring Synergies, Eradicating Poverty." *Journal of Peasant Studies* 36, 1.
Kerblay, Basile. 1985. *Du Mir aux Agrovilles*. Paris: Institut du Monde Sovietique et de l'Europe Centrale et orientale.
___. 1966. "A.V. Chayanov: Life, Career, Works." In Chayanov, *Theory of Peasant Economy*.
Kessel, Joop van. 1990. "Productierituel en technisch betoog bij de Andesvolkeren." *Derde Wereld* 1, 2.
Kinsella, J., P. Bogue, J. Mannion and S. Wilson. 2002. "Cost Reduction for Small-Scale Dairy Farms in County Clare." In Ploeg, Long and Banks (eds.), *Living Countrysides*. Doetinchem, The Netherlands: Elsevier.
Lacroix, A. 1981. *Transformations du Proces de Travail Agricole, Incidences de l'Industrialisation sur les Conditions de Travail Paysannes*. Grenoble, France: INRA.
Lallau, Benoit. 2012. "De la modernité des paysans." *La Découverte/Revue Française de Socio-Economie* 1, 9.
Langthaler, Ernst. 2012. "Balancing Between Autonomy and Dependence: Family Farming and Agrarian Change in Lower Austria, 1945–1980." In Günter Bischof and Fritz Plasser (eds.), *Austrian Lives*. New Orleans: Contemporary Austrian Studies XXI.
Lawner, Lynne. 1975. *Letters from Prison by Antonio Gramsci*. London: Jonathan Cape.
Lenin, Vladimir Illich. 1961 [1906]. "The Agrarian Question and the 'Critics of Marx.'" In *Collected Works, V*. Moscow: Foreign Languages Publishing House.
Li Xiaoyun, Qi Gubo, Tang Lixia, Zhao Lixia, Jin Leshan, Guo Zhanfeng and Wu Jin. 2012. *Agricultural Development in China and Africa: A Comparative Analysis*. London: Routledge.
Lippit, V.D. 1987. *The Economic Development of China*. Arkmont, NY: Sharpe.
Lipton, M. 1977. *Why Poor People Stay Poor: Urban Bias in World Development*. London: Temple Smith.
** Little, Daniel. 1989. *Understanding Peasant China: Case Studies in the Philosophy of Science*. New Haven, CT: Yale University Press.

Long, Norman. 1984. *Family and Work in Rural Societies: Perspectives on Non-Wage Labour*. London: Tavistock.

Long, N., and A. Long. 1992. *Battlefields of Knowledge: The Interlocking of Theory and Practice in Social Research and Development*. London: Routledge.

Luxemburg, Rosa. 1951 [1913]. *The Accumulation of Capital*. London: Routledge.

Maat, Harro, and Dominic Glover. 2012. "Alternative Configurations of Agronomic Experimentation." In J. Sumberg and J. Thompson (eds.), *Contested Agronomy*. London: Routledge.

Mann, S., and J. Dickinson. 1978. "Obstacles to the Development of a Capitalist Agriculture." *Journal of Peasant Studies* 5, 4.

Mariátegui, José Carlos. 1928. *7 Ensayos de interpretación de la realidad Peruana*. Lima: Amauta.

Martinez-Alier, J. 1991. "The Ecological Interpretation of Socio-Economic History: Andean Examples." *Capitalism Nature Socialism* 2, 2.

Marx, Karl. 1963 [1852]. *The Eighteenth Brumaire of Louis Bonaparte*. New York: International Publishers.

____. 1951 [1863]. *Theories of Surplus Value*. London: Lawrence and Wishart.

Marx, Karl, and Friedrich Engels. 1975. *Collected Works, Volume 24*. New York: International Publishers.

Mazoyer, M., and L. Roudart. 2006. *A History of World Agriculture*. London: Routledge.

MDA (Ministério do Desenvolvimento Agrário). 2009. *Agricultura Familiar no Brasil e O Censo Agropecuário 2006*. Brazil: MDA.

Mendras, Henri. 1987. *La Fin des Paysans, suivi d'une reflexion sur la fin des pasans: Vingt Ans Aprés*. Paris: Actes Sud.

____. 1970. *The Vanishing Peasant: Innovation and Change in French Agriculture*. Cambridge: Cambridge University Press.

Milone, P. 2004. Agricoltura in transizione: la forza dei piccoli passi; un analisi neo–istituzionale delle innovazioni contadine. PhD diss., Wageningen University.

Mitchell, T. 2002. *Rule of Experts: Egypt, Techno-Politics, Modernity*. Berkeley: University of California Press.

Moore, Barrington, Jr. 1966. *Social Origins of Dictatorship and Democracy: Lord and Peasant in the Making of the Modern World*. London: Penguin University Books.

Mottura, Giovanni. 1988. "Prefazione A.V. Čajanov: proposte per una possibile linea di lettura di alcuni lavori." In Čajanov, Aleksandr Vasil'evč, *L'economia di lavoro, scritti scelti*. Milan: Franco Angeli/INSO.

Negri, Antonio. 2008. *Reflections on Empire*. Cambridge: Polity Press.
* Netting, Robert. 1993. *Smallholders, Householders: Farming Families and the Ecology of Intensive, Sustainable Agriculture*. Stanford, CA: Stanford University Press.
Norder, Luiz A. Cabello. 2004. *Políticas de Asentamento e Localidade: os desafios da reconstituçao do trabalho rural no Brasil*. Wageningen, The Netherlands: Wageningen University.
Oostindie, Henk. 2013. Multifunctional Agricultural Pathways: Bundles of Resistance, Redesign and Resilience. Wageningen, The Netherlands: Wageningen University.
Oostindie, Henk, Pieter Seuneke, Rudolf van Broekhuizen, Els Hegger and Han Wiskerke. 2011. *Dynamiek en robuutstheid van multifunctionele landbouw, rapportage onderzoeksfase 2: emprisich onderzoek onder 120 multifunctionele landbouwbedrijven*. Wageningen, The Netherlands: LSG Rurale Sociologie, Wageningen University.
Osti, G. 1991. *Gli innovatori della periferia, la figura sociale dell'innovatore nell'agricoltura di montagna*. Torino, Italy: Reverdito Edizioni.
Ostrom, E. 1990. *Governing the Commons: The Evolution of Institutions for Collective Action*. Cambridge: Cambridge University Press.
Paredes, M. 2010. *Peasants, Potatoes and Pesticides: Heterogeneity in the Context of Agricultural Modernization in the Highland Andes of Ecuador*. Wageningen, The Netherlands: Wageningen University.
Paz, R. 2006. "El campesinado en el agro argentino: Repensando el debate teórico o un intento de reconceptualización?" *Revista Europea de Estudios Latinoamericanos y del Caribe* 81.
Perez, Julian. 2012. *A construção social de mecanismos alternativos de mercados no âmbito da Rede Ecovida de Agroecologia*. Paraná, Brazil: MADE-UFPR.
Pérez-Vitoria, Sylvia. 2005. *Les paysans sont de retour, essai*. Arles, France: Actes Sud.
Ploeg, Jan Douwe van der. 2008. *The New Peasantries: Struggles for Autonomy and Sustainability in an Era of Empire and Globalization*. London: Routledge.
___. 2003. *The Virtual Farmer: Past, Present and Future of the Dutch Peasantry*. Assen, The Netherlands: Royal Van Gorcum.
___. 2000. "Revitalizing Agriculture: Farming Economically as Starting Ground for Rural Development." *Sociologia Ruralis* 40, 4.
___. 1990. *Labour, Markets, and Agricultural Production*. Boulder, CO: Westview Press.
Ploeg, J.D. van der, J. Bouma, A. Rip, F. Rijkenberg, F. Ventura and J. Wiskerke. 2004. "On Regimes, Novelties, Niches and Co-production." In J.S.C.

Wiskerke and J.D. van der Ploeg (eds.), *Seeds of Transition: Essays on Novelty Production, Niches and Regimes in Agriculture*. Assen, The Netherlands: Royal van Gorcum.

Ploeg, J.D. van der, A. Long and J. Banks. 2002. *Living Countrysides: Rural Development Processes in Europe — The State of Art*. Doetinchem, The Netherlands: Elsevier.

Ploeg, J.D. van der and Ye Jingzhong. 2010. "Multiple Job Holding in Rural Villages and the Chinese Road to Development." *Journal of Peasant Studies* 37, 3.

Ploeg, J.D. van der, Ye Jingzhong and Sergio Schneider. 2012. "Rural Development Through the Construction of New, Nested Markets: Comparative Perspectives from China, Brazil and the European Union." *Journal of Peasant Studies* 39, 1.

** Polanyi, K. 1957. *The Great Transformation: The Political and Economic Origins of Our Time*. Boston: Beacon Press.

Richards, Paul. 1985. *Indigenous Agricultural Revolution: Ecology and Food Production in West Africa*. London: Unwin Hyman.

Rip, A., and R. Kemp. 1998. "Technological Change." In S. Rayner and E.L. Malone (eds.), *Human Choice and Climate Change*. Vol. 2. Columbus, OH: Battelle Press.

Roep, D. 2000. *Vernieuwend werken; sporen van vermogen en onvermogen (een socio-materiele studie over verniewuing in de landbouw uitgewerkt voor de westelijke veenweidegebieden)*. Wageningen, The Netherlands: Circle for Rural European Studies, Wageningen University.

Rooij, S.J.G. de. 1994. "Work of the Second Order." In Leendert van der Plas and Maria Fonte (eds.), *Rural Gender Studies in Europe*. Assen, The Netherlands: Royal Van Gorcum.

Rosset, Peter Michael, Braulio Machín Sosa, Adilén María Roque Jaime and Dana Rocío Ávila Lozano. 2011. "The Campesino-to-Campesino Agroecology Movement of ANAP in Cuba: Social Process Methodology in the Construction of Sustainable Peasant Agriculture and Food Sovereignty." *Journal of Peasant Studies* 38, 1.

Rosset, Peter Michael, and Maria Elena Martinez-Torres. 2012. "Rural Social Movements and Agroecology: Context, Theory, and Process." *Ecology and Society* 17, 3.

Sabourin, E. 2006. "Praticas sociais, políticas públicas e valores humanos." In S. Schneider (ed.), *A Diversidade da Agricultura Familiar*. Porto Alegre, Italy: UFRGS Editora.

Saccomandi, V. 1998. *Agricultural Market Economics: A Neo-Institutional Analysis of Exchange, Circulation and Distribution of Agricultural Products*.

Assen, The Netherlands: Royal van Gorcum.
Salas, Maria, and Hermann Tilmann. 1990. "Andean Agriculture — A Development Path for Peru?" In ILEA newsletter, March.
Salter, W.E.G. 1966. *Productivity and Technical Change*. Cambridge: Cambridge University Press.
Savarese, E. 2012. *Young People's Perception of Rural Areas: A European Survey Carried Out in Eight Member States*. Rome: Rete Rurale, Ministero delle Politiche Agricoli, Alimentari e Forestali.
*Schneider, S., and P. Niederle. 2010. "Resistance Strategies and Diversification of Rural Livelihoods: The Construction of Autonomy among Brazilian Family Farmers." *Journal of Peasant Studies* 37, 2.
Schneider, S., S. Shiki and W. Belik. 2010. "Rural Development in Brazil: Overcoming Inequalities and Building New Markets." *Rivista di Economia Agraria* LXV, 2
Schutter, Olivier de. 2011. "How Not to Think of Land-Grabbing: Three Critiques of Large-Scale Investments in Farmland." *Journal of Peasant Studies* 38, 2.
** Scott, James C. 2009. *The Art of Not Being Governed: An Anarchist History of Upland Southeast Asia*. New Haven, CT: Yale University Press.
** ___. 1998. *Seeing Like a State: How Certain Schemes to Improve the Human Condition Have Failed*. New Haven, CT: Yale University Press.
___. 1976. *The Moral Economy of the Peasant*. New Haven, CT: Yale University Press.
Sender, J., and D. Johnston. 2004. "Searching for a Weapon of Mass Production in Rural Africa: Unconvincing Arguments for Land Reform." *Journal of Agrarian Change* 4, 1 & 2.
Sennett, R. 2008. *The Craftsman*. New Haven, CT: Yale University Press.
Sevilla Guzman, Eduardo. 1990. "Redescubriendo a Chayanov: hacia un neopopulismo ecológico." *Agricultura y Sociedad* 55.
Sevilla Guzman, Eduardo, and Manuel González de Molina. 2005. *Sobre a evolução do conceito de campesinato*. Brasília: Via Campesina do Brasil/Expressão Popular.
** Shanin, Teodor. 2009. "Chayanov's Treble Death and Tenuous Resurrection: An Essay About Understanding, About Roots of Plausibility and About Rural Russia." *Journal of Peasant Studies* 36, 1.
___. 1986. "Chayanov's Message: Illuminations, Miscomprehensions, and the Contemporary 'Development Theory.'" Introduction to A.V. Chayanov, *The Theory of Peasant Economy*. Madison: University of Wisconsin Press.
Slicher van Bath, B.H. 1978. "Over boerenvrijheid (inaugurele rede

Groningen, 1948)." In B.H. Slicher van Bath and A.C. van Oss (eds.), *Geschiedenis van Maatschappij en Cultuur*. Baarn, The Netherlands: Basisboeken Ambo.

____. 1960. *De agrarische geschiedenis van West–Europa, 500–1850*. Utrecht/ Antwerpen, The Netherlands: Het Spectrum.

Sonneveld, M.P.W. 2004. "Impressions of Interactions: Land as a Dynamic Result of Co-Production between Man and Nature." PhD diss., Wageningen University.

Sperotto, F. 1988. "Aproximación a la vida y a la obra de Chayanov." *Agricultura y Sociedad* 48

Spoor, Max. 2012. "Agrarian Reform and Transition: What Can We Learn From 'The East'?" *Journal of Peasant Studies* 39, 1.

Steenhuijsen Piters, B. de. 1995. *Diversity of Fields and Farmers: Explaining Yield Variations in Northern Cameroon*. Wageningen, The Netherlands: Agricultural University.

Stoop, Willem A. 2011. "The Scientific Case for System of Rice Intensification and its Relevance for Sustainable Crop Intensification." *International Journal of Agricultural Sustainability* 9, 3.

Sumberg, J., and C. Okali. 1997. Farmers' Experiments: Creating Local Knowledge. Boulder, CO: Lynne Rienner Publishers.

Sumberg, James, and John Thompson (ed.). 2012. *Contested Agronomy: Agricultural Research in a Changing World*. London: Routledge.

Sumberg, James, John Thompson and Philip Woodhouse. 2013. "Why Agronomy in the Developing World Has Become Contentious." *Agriculture and Human Values* 30, 1.

Thiesenhuisen, W.C. 1995. *Broken Promises: Agrarian Reform and the Latin American Campesino*. Boulder, CO: Westview Press.

* Thorner, D. 1966. "Chayanov's Concept of Peasant Economy." In Chayanov, *Theory of Peasant Economy*.

Timmer, C.P. 1970. "On Measuring Technical Efficiency." *Food Research Institute Studies in Agricultural Economics, Trade and Development* 9, 2.

Timmer, W.J. 1949. *Totale Landbouwwetenschap, een cultuurphiloophische beschouwing over landbouw en landbouwwetenschap als mogelijke basis voor vernieuwing van het landbouwkundig hoger onderwijs*. Groningen, The Netherlands: Wolters.

Toledo, Victor M. 2011. "La agrocología en Latinoamercia: tres revoluciones, una misma transformación." *Agroecología* 6

* ____. 1990. "The Ecological Rationality of Peasant Production." In M. Altieri, *Agroecology and Small Farm Development*. Ann Arbor, MI: CRC Press.

Vanloqueren, G., and P.V. Baret. 2009. "How Agricultural Research Systems

Shape a Technological Regime that Develops Genetic Engineering but Locks Out Agroecological Innovations." *Research Policy*, 38.

Veltmeyer, H.. 1997. "New Social Movements in Latin America: The Dynamics of Class and Identity." *Journal of Peasant Studies* 25, 1.

Ventura, F. 2001. "Organizzarsi per Sopravvivere: Un analisi neo-istituzionale dello sviluppo endogeno nell'agricoltura Umbra." PhD diss., Wageningen University.

___. 1995. "Styles of Beef Cattle Breeding and Resource Use Efficiency in Umbria." In J.D. van der Ploeg and G. van Dijk (eds.), *Beyond Modernization: The Impact of Endogenous Rural Development*. Assen, The Netherlands: Royal Van Gorcum.

Vera Delgado, J. 2011. "The Ethno-Politics of Water Security: Contestations of Ethnicity and Gender in Strategies to Control Water in the Andes of Peru." Wageningen, The Netherlands: Wageningen University.

Vijverberg, A.J. 1996. *Glastuinbouw in ontwikkeling, beschouwingen over de sector en de beinvloeding ervan door de wetenschap*. Delft, The Netherlands: Eburon.

Visser, Jozef. 2010. "Down to Earth: A Historical-Sociological Analysis of the Rise and Fall of 'Industrial' Agriculture and the Prospects for the Re-rooting of Agriculture from the Factory to the Local Farmer and Ecology." PhD diss., Wageningen University.

Vitali, S., J.B. Glattfelder and S. Battiston. 2011. "The Network of Global Corporate Control." <arxiv.org/abs/1107.5728v1>.

Vlaslos, Stephen. 1986. *Peasant Protests and Uprisings in Tokugawa, Japan*. Berkely: University of California Press.

Vries, Egbert de. 1948. *De Aarde Betaalt: de rijkdommen der aarde en hun betekenis voor de wereldhuishouding en politiek*. Den Haag, The Netherlands: Uitgeverij Albani.

___. 1931. *De landbouw en de welvaart in het regentschap Pasoeroean, bijdrage tot de kennis van de sociale economie van Java*. Wageningen, The Netherlands: Landbouwhogeschool.

Wanderley, Maria de Nazareth Baudel. 2009. "Em busca da modernidade social: uma homenagem a Alexander V. Chayanov." In Maria Wanderley, *O mundo rural como um espaço de vida; reflexões sobre a propriedade da terra, agricultura familiar e ruralidade*. Porto Alegre, Brazil: PGDR/UFRGS Editora.

Warman, A. 1976. *Y venimos a contradecir, los campesinos de Morelos y el Estado Nacional*. Mexico City: Ediciones de la Casa Chata.

Wartena, D. 2006. "Styles of Making a Living and Ecological Change on the Fon and Adja Plateaux in South Benin, ca. 1600–1900." PhD diss.,

Wageningen University.
Weis, Tony. 2010. "The Accelerating Biophysical Contradictions of Industrial Capitalist Agriculture." *Journal of Agrarian Change* 10, 3.
** ___. 2007. *The Global Food Economy: The Battle for the Future of Farming.* London: Zed Books.
White, Ben. 2011. *Who Will Own the Countryside? Dispossession, Rural Youth and the Future of Farming.* The Hague: International Institute of Social Studies.
Wiskerke, J.S.C., and J.D. van der Ploeg. 2004. *Seeds of Transition: Essays on Novelty Production, Niches and Regimes in Agriculture.* Assen, The Netherlands: Royal Van Gorcum.
Wit, C.T. de. 1992. "Resource Use Efficiency in Agriculture." *Agricultural Systems* 40.
Wolf, Eric R. 1969. *Peasant Wars of the Twentieth Century.* New York: Harper and Row.
Woodhouse, Philip. 2010. "Beyond Industrial Agriculture? Some Questions about Farm Size, Productivity and Sustainability." *Journal of Agrarian Change* 10, 3.
Wu Xiang. 1998. "The Tortuous Processes of Rural Reform." *The Century* 3.
Yang, M.C. 1945. *A Chinese Village: Taitou, Shantung Province.* New York: Columbia University Press.
Ye Jingzhong. 2002. *Processes of Enlightenment: Farmer Initiatives in Rural Development in China.* Wageningen, The Netherlands: Wageningen University.
Ye Jingzhong, Rao Jing and Wu Huifang. 2010. "Crossing the River by Feeling the Stones: Rural Development in China." *Rivista di Economia Agraria* 65, 2.
* Ye Jingzhong, Wang Yihuan and Norman Long. 2009. "Farmer Initiatives and Livelihood Diversification: From the Collective to a Market Economy in Rural China." *Journal of Agrarian Change* 9, 2.
Yingfeng Xu. 1999. "Agricultural Productivity in China." *China Economic Review* 10.
Yong Zhao, and J.D. van der Ploeg. 2009. "Telling Data: An Analysis of the Note Book of a Chinese Farmer." *Journal of China Agricultural University* 26, 3.
Yotopoulos, P.A. 1974. "Rationality, Efficiency and Organizational Behaviour Through the Production Function: Darkly." *Food Research Institute Studies* 13, 3.
Zanden, J.L. van. 1985. *De Economische Ontwikkeling van de Nederlandse Landbouw in de Negentiende Eeuw, 1800–1914.* Wageningen, The Netherlands: AAG Bijdragen, Landbouwuniversiteit.

Index

Agrarian Revolution, in Britain, 112–113
Agrarian sciences, 5, 17
   Food-engineering, 53–55, 64, 116–119, 123
   Hegemonic bias of, 110, 112, 114, 117–118, 122–123,
   Impact on social agronomy, 5, 110–111, 117
   Pitfalls of, 5, 23
   Role in entrepreneurial farming, 53–55, 100–101, 114, 117–119
   Role in intensification of farming, 110–118
   Soil chemistry, 53, 112–115
Agriculture
   Corporate control of, 67, 75, 82–84, 118, 130
   Entrepreneurial, 53–56, 58, 64–66, 75–76, 128
   Evolution of, 55, 65, 68–74
   Global requirements of, 5
   Large scale intensive, 64
   Market driven, 55–56, 58
   Organic, 54, 100–101, 129
   Policy and subsidization, 63–64
   Relation of producers and external forces, 56–60, 66–67, 76, 78–87, 118
   Role in development, 4, 5, 13, 119–123
   Role in healing the environment, 13, 121
   Sustainability of, 4, 28, 54–58, 121
   See also, Peasant unit of production; Yields
Agro-ecology, 100, 117–118, 123, 127
Agro-industries, 55, 75, 82–84, 112, 130
Art of farming
   Book by L. Columella, 22
   Practice of, 6–11, 48–49, 69–70, 117–118, 123

Balances in peasant farming
   Between agents involved in agriculture, 6–11, 20, 23
   Between agrarian and demographic growth, 47, 87–88, 109
   Between autonomy and dependence, 60–62, 76, 79–80, 89
   Between capital formation and labour, 30, 37, 44–47
   Between internal and external resources, 56–60, 66–67, 76, 78–87, 118
   Between labour and consumption, 5–10, 22–24, 28, 33–37, 66, 128
   Between members of the farming community, 27–28
   Between people and nature, 7–8, 10, 48–54
   Between production and reproduction, 28, 42, 54–58, 70, 76–77
   Between scale and intensity, 63–66, 84–85, 103–105
   Between town and countryside, 78–82, 124, 127
   Between utility and drudgery, 6, 29–34, 37–46, 59, 128–131
   Other balances, 77
   Political relevance of, 34–37
   Synthesis of peasant farming, 68–73
   See also, Peasant unit of production
Benvenuti, Bruno, 55, 83
Bernstein, Henry, 2, 13, 17, 70, 117
Bizzozzero, Antonio, 110
Boserup, Ester, 87
Bolshevik policy toward peasants, 19–21, 46, 79
Bolivia, farmers of, 119
Bové, José, 130
Brazil
   Land reform in, 3, 12–14
   Landless Workers' Movement, 127
   Legal status of peasantry, 123
   Legas Camponesas, 3, 123

Market commons, 68, 130
Peasant labour movements, 3, 78, 123, 127
Repeasantization 12, 14, 127

Campesinistas, 3
Capital formation
  Alternative sources, 14–17, 24–26, 57
  As practised by peasants, 30, 37–47, 54–60, 119
  As reproduction, 28, 42, 54–58, 70
  Calculation of, 102–103
  Conflicts of interest in, 9–10
  Impact on prices of farm produce, 15, 75, 79
  Role of agriculture in, 4
Capitalism
  As applied to farming, 31–32, 53–56, 71, 82–84, 101–103
  Comparison with peasant economics, 29–34, 44–46, 55, 76–77, 102–103
  Crisis of, 11
  Impact on peasant families, 15, 29, 59, 107–108, 119, 124
  Peasant alternatives to, 13–16, 24–26, 28–29, 76–77, 121–123
Capitalist agriculture
  In comparison with peasant farming, 24–32, 36–37, 71–73, 76–77, 119–123
  Investments of, 5, 56, 119–120
  Logic of, 24, 29, 31, 71–73, 90–91
  Marxist view of, 36–37, 74, 77
  Peasant view of, 5, 11, 14–17, 24–26
  Production methods of, 103, 110–119
  Recessions of, 75
  Resistance to, 14–15, 62, 84–85, 88, 127
  Rise of, 1, 75–76
  Style of, 53–56, 58, 64–66, 75–76, 128
  Wastefulness of, 53–56, 119–120
  Weaknesses of, 16, 31, 75, 90, 119–121

Chayanov, Alexander
  "Genius" of, 17–21
  Influence on other thinkers, 20–21
  On capitalism in relation to peasants, 15–17, 24–26, 31–32, 44, 76
  On commoditization, 59, 83
  On empirical research into peasant issues, 12–13, 16–18, 22
  On peasant "self-exploitation," 42–44, 47
  On peasant coproduction, 63, 83–84
  On peasant labour and return, 33–34, 37–47, 59–60, 90, 109
  On peasant land use, 31, 33–35
  On the future of peasants in Russia, 17, 78–79
  On the needs of peasants, 34–42, 61, 86, 108–109
  Micro-level analysis of peasant farms, 17, 23, 43–44
  On social agronomy, 9–11, 18, 54, 65
  On the peasant unit of production, 5–11, 15–16, 19, 24–26
  On the potential of peasant farmers, 6, 17–19, 22, 37, 41, 119–120
  On the social identity of peasants, 1, 3, 6, 17–18, 74
  Writings of
    Economy of Labour, 51–52
    Essay About the Functioning of the Peasant Farm, 90
    The Journey of my Brother Alexis to the Land of Peasant Utopia, 78
    Theory of the Peasant Economy, 74, 89
Chemicalization of agriculture, 52–53, 112–115, 118, 123–124
China
  Commercialization in, 28, 59, 87, 130
  Commons, 14, 68
  Household Responsibility System (HRS), 125–127

INDEX

Innovative markets, 130
Migration of labour, 81–82
Novelty production, 99
Population-resources imbalance, 87
Reforms after 1949, 2–3
Reforms after 1978, 3–4, 35, 99, 125–127, 131
Relation to the land, 51
Repeasantization, 4, 12, 35, 81–82, 125–127, 131
Surplus extraction, 61, 125–126
Class
    Diversification of, 13–14, 73–77, 82
    Identity of the peasantry, 2, 13–14, 19–20, 36–37, 61
    Role in communes, 83
    Solidarity among peasants, 88
Columella, Luciano, 22
Commoditization, degree of, 46, 55–56, 58–60, 67, 76–77
Commodity relations
    Between producers, processors and marketers, 79–80, 82–84, 88
    Concerning rice in Guinea Bissau, 26–28
    For market-driven agriculture, 46, 55–56, 58–59, 77
    For peasant families, 15, 26, 33, 35–36, 71–73
    Impact of low commodity prices, 32, 56
    In new cooperative structures, 80–81, 84
    Noncommodities, 50–51, 72–73, 118
    Of Italian dairy farmers, 50–51, 121–122, 124
Common Agricultural Policy (of the EU), 65, 86
Commons
    Shared investments, 14
    Shared labour, 30, 83–85
    Shared markets, 14, 68, 84, 130
    Shared natural resources, 14, 22, 68, 84–85, 89–90

Shared produce, 27–29
Communes
    Community run, 20, 52, 83–85
    Kolkhozes, 2, 35
    State run, 2, 125–126
    Vertical (as opposed to horizontal cooperation), 19, 35
Competitiveness
    Of corporate agribusiness, 75–76, 128
    Of peasant farmers, 16–17, 71, 119–123, 129–130
Cooperatives
    Labour, 30, 104–105
    Machine, 30
    Market-controlled, 84, 104
    Movements of, 19, 80, 83–85
    Production of, 84–85, 104, 125–126
    State-controlled, 2, 35, 85, 125–126
    Vertical, 19
Coproduction
    Evolution of, 76
    Shared machinery, 30
    With Nature, 48–49, 50–54, 63, 121, 123, 129
Craftmanship, 41–42, 50, 97
Credit policies, 56, 67, 129, 135
Crises
    Environmental, 52
    Of agriculture, 75, 86, 128
    Of capitalism, 11, 67, 128
Cuba, 2–3

Descampesinistas, 3
Decommoditization, 60
Differentiation
    Among classes, 62
    Among peasants, 73–77
Distribution of wealth
    Among farmers, 28–29, 71, 86
    In society, 60–61, 67, 82, 93, 108
Diversification (or multifunctionality), 120, 129–130
Durrenberger, Paul, 11, 77
Dutch peasant farming. *See* Netherlands, peasant farming.

Economy of Labour, by A. Chayanov, 51–52
Empoverishment. *See* Involution.
Equilibria in farming
   In agricultural policy, 86
   In farm management, 7–10, 22, 34–38, 43–44, 59
   Resilience of, 62
   See also Balances in peasant farming.
Essay About the Functioning of the Peasant Farm, by A. Chayanov, 90
European Union
   Farming policies, 65, 86
   Repeasantization in, 127–131
Exchange relations. *See* Commodity relations.

Family farm. *See* Peasant unit of production.
Farming styles
   Economical farmers, 30, 63–64
   Evolution of, 55, 65, 68–74
   Hybridity of, 66
   Labour intensive, 72, 85, 90–96, 103–107, 109–110
   Labour saving, 64, 73, 84
   Large-scale, entrepreneurial, 53–55, 64–66, 75–76, 128
   Mutual help, 30, 77
   Vangard farmers, 30
   See also Peasant unit of production.
Fei Xiao Tung, 74
Fertilizer, chemical, 52–53, 112–115, 118, 123–124
Fertilizer, organic, 57, 100–101, 118
Fine tuning
   Of farm environment, 10, 69, 95
   Of farming skills, 52, 98–101
   Of technical efficiency, 98
Flows of farming processes
   Of farm renewal, 68, 100
   Of machinery, 30
   Of produce, 26–29, 57–58, 72
   Of resources, 57–58, 82, 118
Food and Agricultural Organization (FAO), 5, 128

Food empires,
   Economic power of, 67, 70, 79, 82, 117, 122
   Impact on peasants, 67, 75, 82–84, 118, 130
Food supplies
   Production solutions by peasants, 13, 70, 119–123, 125–127
   Scarcity of, 4–5, 67
Friedmann, Harriet, 75

Gender relations
   In peasant production, 9, 35, 42–43, 72, 93
Marriagability of peasants, 108
Genetically modified organisms (GMOs), 53, 118
Granary management, 27–29
Green Revolution, 53, 75, 101, 116–117, 123
Growth factors, or limiting factors, 95–96, 99, 115, 123
Guinea Bissau
   Land reform in, 3
   Rice farm management, 26–30, 56, 108

Hardt, Michael, and Negri, A., 13, 22
Hayami, Yujiro, and Ruttan, V., 63, 94, 109
Hobsbawm, Eric, 77
Household income (for farmers), 32–34, 37–38, 67
   See also Labour; Peasant unit of production

Intensification of farming
   In peasant farming, 89–92, 101–107
   In response to land scarcity, 33, 71–72, 89, 109
   In response to price-cost squeeze, 109–110
   Labour driven, 72, 85, 90–96, 103–107, 109–110
   Limits to, 105–107
   Through agrarian sciences, 110–119

Through development of marginal land, 90–91
Through improvements in resources, 90, 92–95, 98–99, 112–113, 118
Through novelty production, 99–100, 111–112, 117, 120–121
Through technical or mechanical means, 53–55, 64–65, 92–93, 98–105, 112
Intensity of resources use
   In relation to scale, 63–66, 84–85, 91, 94–95, 103–105
   Of capital investment, 37, 64–65, 90–91, 101–105
   Of cultivation, 33, 64, 90, 103–105
   Of diversification, 32–35, 127–131
   Of labour process, 37–42, 52, 63, 72, 85
   Of skill, 93, 95–105
Inverted supply curve, 23
Involution
   As deactivation or stagnation of peasant agriculture, 70, 105–108
   Due to competition from large-scale agribusiness, 107–108
   Due to impoverishment, 108
   Due to population/resources disparity, 87–88, 108
Interstices between structures, 14–16, 69, 84, 130
"Inverse relationship" of farm size to productivity, 64, 91, 103
   See also Scale of farming.
Italy
   Dairy farming in, 50–51, 121–122, 124
   Diversity of farm vocations, 32
   land reform in, 3–4
   peasant exchange mechanisms, 29
   Peasant management methods, 41–42, 124
   Peasant relation to living nature, 50–51, 121–122
   role of peasant movements in, 21
   Southern question, 4, 21

Urban-rural migration, 81

Journey of my Brother Alexis to the Land of Peasant Utopia, The, by A. Chayanov, 78
Kautsky, Karl, 15–16, 45–46
Knowledge in farming
   As "subjective evaluation," 43–44, 98
   As local tradition, 22, 24–25, 97–105, 112–113, 117
   As skill-oriented technology, 93, 95–105, 123
   From direct experience, 35–36, 50, 55, 97, 120
   From peasant innovation, 69, 76, 95–97, 99–100, 111–112
   Local vs. scientific, 53, 97, 114–118
Kolkhozes, 2, 35

Labour
   As investment, 24–25, 63, 72
   "Black" immigrant, 128
   Conditions for peasants, 14–16, 24–26, 31–34, 37–42
   In relation to consumption, 5–10, 22–24, 28, 32–37, 66, 128
   Justice in compensation, 34–35, 61
   "Self-exploitation," 44–46
   Unions for farm workers, 85
   Utility vs. drudgery, 38–42, 67
   Wages or return for peasants, 6, 29–34, 37–46, 59, 126–131
   See also Wages, or Labour Income
Land occupations, 12, 85
Land reform, or redistribution, 3–4, 71, 108
Langthaler, Ernst, 62, 66
Law of diminishing returns, 101, 105–107
League for Agrarian Reform (in Russia), 3
Lenin, Vladimir
   Views on land reform, 3, 31
   Views on the peasant question, 1, 16, 21, 45, 106–107
Liebig, Justus von, 112, 114–115, 123

Limiting factors, or growth factors, 95–96, 99, 115
Little, Daniel, 16, 60–61, 75, 88
Long, Norman, and Long, A., 43, 77
Luxemburg, Rosa, 36–37

Mariátegui, José Carlos, 3–4, 16
Markets
    Alternative exchange mechanisms, 14–16, 24–26, 29–30, 57–58
    Dependency on, 55–56, 58–62, 66, 76–77, 79–80
    Downstream, 67, 79, 82–83
    Global trends of, 79–80
    Globalization of, 67, 75–76, 79–80
    Impact of cheap imports and subsidies, 29, 56
    Impact on nature and sustainability, 50–51, 55–56, 67
    Machinery flows, 30, 57
    Nested markets, 30, 130
    Of processing industries, 67, 82–83
    Peasant interaction with, 10–11, 14, 24–25, 33, 43, 57–58
    Prices for farm produce, 15–16, 45–46, 55–56, 66–69, 79
    Squeeze on agriculture, 66–69, 79–80, 109
    Upstream, 67, 83
    Volatility of, 120
Marketing chains, 82–83
Martinez-Alier, J., 21
Marx, Karl, 17–18, 20, 32, 36, 74
Marxist theory concerning peasants, 19–22, 31, 36–37, 74–75, 106–109
Mazoyer, M., and Roudart, L.
    On crisis of capitalism, 11, 86
    On history of peasant farming, 89, 111–112, 114
Mechanization of farming, 52–53, 104, 112, 114–116, 124
Mendras, Henri, 14, 96
Mexican farmers, 2–3, 75
Micro-level farm analysis, 17, 23–26, 33–34, 49, 87
    See also Guinea Bissau, Rice farm management.
Migration
    As town-country balance, 81–82, 84, 127
    Away from farm families, 108
    To cities, 56, 81–82
Moral economy, 40, 54, 77
Mottura, Giovani, 11, 88
Multifunctionality (or diversification), 120, 129–130
Multitude of peasant diversity, 14

Narodniki, 1, 17, 21
Negri, Antonio, 13, 22, 130
Neoclassical economics, 76–77
Neoliberalism, 80
Nested markets, 14, 30, 129–130
Netherlands, farmers of, 7, 29, 111, 124, 129–130
Netting, Robert
    Influence by Chayanov, 21
    On declines of peasant farming, 87, 107
    On peasant farming methods, 47, 114, 121, 125
    On repeasantization, 35, 68, 75
Noncommodities, 50–51, 72–73, 118
Norway farmers, 82
Novelty production, 99–101, 111–112, 117, 120–121, 123

Patrimony,
    As "family capital," 25–26, 29–30
    In relation to markets, 67
    In the Mediterranean, 32
Peasant communities
    Commons of, 14, 19, 22, 68
    In Russia, 17, 19
    Mutual help, 30
    Role of, 3, 19
    Tensions within, 9
Peasant movements
    Agro-ecological, 54, 60, 76, 117–118, 123
    Alliances of, 4, 20, 21
    Contributions and struggles of, 10, 13–15, 35, 61, 88
    Farming as resistance, 68–69, 88

In China, 35, 125–127
In Europe and Russia 17, 21, 47, 127–131
Land occupation, 85, 127
Repression of, 123
Transnational agrarian movements (TAMs), 13, 20
"Peasant question," or agrarian question
  Attitudes toward the peasantry, 1–5, 11, 16, 36–37, 45
  Concerning peasant autonomy, 108–109
  Concerning scales of efficiency, 31, 33–35, 84–85
  Critical theory on, 11–14
  In Russia, 1–3, 21–22
  In the developing world, 2–4, 22, 108
  In Western Europe, 2, 21
  Marxist thought on, 20–21, 31, 36–37, 74–75, 106–109
  Political relevance of, 1–2, 5, 11–15, 34–35, 119–123
  Preobrazensky-Bucharin debate, 22
  See also Chayanov, Alexander.
Peasant unit of production
  Autonomy of, 34–36, 42, 58–65, 72–76, 89
  Balance with reproduction, 28, 42, 54–58, 70, 76–77
  Balances maintained, 5–10, 22–24, 33–42, 68–73, 77. *See also* Balances in peasant farming.
  Comparison with capitalist production, 24–32, 36–37, 71–73, 76–77, 119–123
  Competitive power of, 16–17, 71, 119–123, 129–130
  Consumption in, 510, 22–24, 28–37, 66, 128
  Contributions of, 6, 13, 22, 70, 119–123
  Coproduction in, 63, 84–85, 72
  Cycles of, 72, 74
  Exchange mechanisms, 29–30, 56–60
  Family and farm economics, 24–26, 32–44, 47, 87–88, 109
  Intensity and scale in, 63–66, 84–85, 103–105
  Interactions with outside agents, 6–11, 15, 20, 23, 78–87
  Labour in, 6, 29–34, 37–46, 59, 128–131
  Mature form of, 68–73
  Micro-level analysis of, 17, 23–26, 43–46
  Multiplicity of activities, 32–36, 65–66, 68–70, 129–130
  Mutual help and reciprocity, 30
  Organizational plan of, 10–11, 14, 22, 34–46, 63
  Political relevance of, 34–37
  Relation to capital formation, 14–17, 24–26, 30, 37, 44–47
  Relation to living nature, 78, 10, 48–57, 92
  Relation to market forces, 10–11, 15–16, 24–26, 32–33, 71
  Relation to urban centres, 78–82, 124, 127
  Resilience of, 17, 66, 68–69, 120–121
  Role in processing and marketing, 82–84, 129
  "Subjective evaluation" in, 42–44, 47
  Sustainability of, 41–42, 54–58, 71, 120–121
  Value added, 70–71, 73
  Yields of. *See* Yields.
  See also Labour; Wages and labour income.
Peasants
  Class status of, 2, 13, 14, 19–21, 36–37, 61–62
  Coalitions of, 2, 21
  Diversity of, 14, 73–77
  Economic power of, 1–6, 22, 35, 125–131
  Modern needs of, 34–35, 66
  Oppression of, 6, 29, 61, 70
  Poverty of, 4, 11, 67–68
  Relation to the state, 34–35, 84–87
  Resources of, 56–60, 66–67, 76,

78–87, 118
  Working conditions of, 14–16,
    24–26, 31–34, 37–42
Peru, farmers of
  Agricultural cooperatives, 84–85
  Balance with nature, 51
  Credit and investment policies,
    104
  Indigenous peoples' issues, 4
  Land reform in, 3
  Peasant production methods,
    103–105, 113, 119
  Relation to water, 92
Perez, Julian, 105, 130
Polanyi type of "anti-market device,"
  10
Portugal, farming in, 3, 119
Preobrazensky-Bucharin debate, 22

Regression of peasant agriculture. *See*
  Involution.
Repeasantization, 2, 6, 12, 16, 35,
  125–131
Resources
  Exchange of by peasants, 56–58
  Internal or external, 56–60, 66–67,
    76, 78–87, 118
  Reproduced, or regenerated,
    54–58, 98–99, 112–115
  Shared, or common, 14, 22, 68,
    84–85, 89–90
Resource base
  As family capital, 25–26, 29–30,
    62, 67, 73
  As unity of social and natural
    forces, 72
  Available to peasants, 71, 80
  Depletion or expropriation of, 56,
    71
  Development by peasants, 39, 55,
    71–72, 98–99
Roep, Dirk, 7, 22
Russia
  Land redistribution, traditional, 3,
    89
  Modern farming conditions, 66
  Traditional agriculture in, 1,
    17–18, 33, 52, 66

Zemstov land statistics, 89
Russian revolution
  Bolshevik policy toward the
    peasants, 19–21, 46, 79
  Role of cooperatives in, 3, 19–20,
    123
  Role of peasants in, 1–3, 22, 25, 75
  Social Revolutionary Party policy,
    21

Scale of farming
  Appropriate production units,
    31–34, 52, 63–66, 84–85,
    103–105
  Bias for larger farm units, 31,
    64–65, 94
  Farm size/productivity
    relationship, 63–66, 84–85, 91,
    94, 103–105
Scott, James
  On autonomy for peasants, 14, 86,
    117
  On the moral economy, 40
Second Agricultural revolution of
  modern times, 124
"Self exploitation" of the peasantry,
  44–46
Sevilla Guzman, Eduardo, 17, 21
Shanin, Teodor, 5, 17, 44
Slicher van Bath, B.H., 62, 89
Social Agronomy
  As book, by A. Chayanov, 6, 18,
    34, 49, 65, 110
  As coproduction, 54, 65, 123
  As the art of farming, 6–11, 48–49,
    69–70, 117–118, 123
  In the peasant production unit, 18,
    34–35, 65, 123
  In relation to other theories, 110,
    117, 123
Squeeze on agriculture, 4, 67, 79–80,
  109, 129
SRI (System of Rice Intensification),
  100–101, 105, 123
Stagnation of peasant agriculture. *See*
  Involution.
State policies
  In relation to peasants, 84–87,

125–126
  Of the Bolsheviks, 19–21, 46, 79
  On subsidies for agriculture, 86
Styles of farming. *See* Farming styles.
"Subjective evaluation" in peasant farming, 8, 42–44, 47
Subsidies for agriculture, 86
Surplus extraction from peasants, 46, 60–62, 70, 85–86

Theory of the Peasant Economy, by A. Chayanov, 74, 89
Timmer, W.J., 52, 123
Toledo, Victor, 50–51, 127
Transnational agrarian movements (TAMs), 13, 20

Urban bias, 56, 79

Vertical cooperation (as opposed to horizontal integration of farms), 19
Via Campesina, 13, 20, 127–128
Vietnam
  Land reform in, 2–3
  Peasant role in revolution, 22
  Repeasantiztion, 12
Vries, Egbert de, 4, 21, 52, 123
Wages, or labour income
  Nonwage peasant income, 24–32, 37, 47, 59–60, 70
  Profitability of, 90–91, 101–105, 109–110, 126–131
  See also Labour

Yields
  As affected by agrarian science, 110–118
  As affected by labour intensity, 72, 85, 90–96, 103–105, 109
  As affected by mechanization, 103–105, 111, 114
  As affected by social relations, 89, 93–94, 97, 125–126
  As affected by state policy, 84–85, 125–126.
  Balance and Sustainability of, 41–42, 54–58, 71, 120–121
  Benefits of, 89–91
  From bioengineering, 116–117
  From diversification, 130
  From novelty production, 99–100, 130
  From repeasantization, 125–127, 129, 131
  Intensification of, 89–92, 101–107
  Limiting or growth factors of, 95–96, 99, 115
  Of peasant as opposed to capitalist farms, 84–86, 121–123, 100–105, 129–130
  Potential for ever-increasing returns, 101, 105–107, 119–123
  See also Intensification of farming; Intensity of resources applied
Youth
  Migration to cities, 56, 81
  Roles in peasant farming, 9

Zemstov statistics, 18, 89

www.ingramcontent.com/pod-product-compliance
Ingram Content Group UK Ltd.
Pitfield, Milton Keynes, MK11 3LW, UK
UKHW021838140426
5217IPUK00022B/1507